AIRCRAFT MAINTENANCE AND SERVICE

DRAKE'S AIRCRAFT MECHANIC SERIES

AIRCRAFT WOODWORK

AIRCRAFT WELDING

AIRCRAFT SHEET METAL

AIRCRAFT ENGINES

AIRCRAFT ELECTRICAL SYSTEMS,
      HYDRAULIC SYSTEMS,
      AND INSTRUMENTS

AIRCRAFT MAINTENANCE AND SERVICE

AIRCRAFT ENGINE MAINTENANCE AND SERVICE

AIRCRAFT MAINTENANCE AND SERVICE

by COLONEL ROLLEN H. DRAKE, B. S., M. A.

RESEARCH ENGINEER, PILOT AND GROUND SCHOOL INSTRUCTOR
FORMERLY: INSTRUCTOR, VOCATIONAL AND AVIATION SUBJECTS, LOS
ANGELES CITY SCHOOLS; CHIEF, AIR AGENCY UNIT AND CIVILIAN PILOT
TRAINING SPECIALIST, CIVIL AERONAUTICS ADMINISTRATION; SUPPLY
SPECIALIST, OFFICE OF THE QUARTERMASTER GENERAL

Originally published in 1949

ISBN: 978-1-940001-39-5

The Aviation Collection
by
Sportsman's Vintage Press
2015

This book is dedicated to
MY BELOVED WIFE
*without whose inspiration and help
this book could not have been written*

PREFACE

THE airplane, like the human body, is subject to many little ills. Unlike the human body, however, the airplane is not able to get well by itself. The airplane, no matter how slight its damage, if left to itself gets worse until it leads to the complete failure of the affected part.

The field of aircraft maintenance and service is broad and takes in all of the mechanical trades. In this field, the woodworker, welder, sheet-metal worker, machinist, electrician, instrument specialist, and the all-around certificated mechanic, find their places. Maintenance and service problems of large modern aircrafts require men who are expert in many fields.

On the subject of woodwork, the different kinds of woods used in aircraft structure and their use are thoroughly discussed. There are many practical problems in construction which explain how approved repairs are to be made. Metals used in aircraft structures are also discussed, and many practical illustrations and explanations of the approved manner of making metal repairs are included.

This book is written to bring before the layman, the student, the teacher and the certificated aircraft mechanic the fundamentals of aircraft maintenance and service. Each branch of the field of maintenance and service is adequately covered. This book not only explains what operations to perform, but also goes thoroughly into the field of inspection and tells how each operation should be performed. Perhaps the most outstanding feature of this book is the non-technical, simple language in which it is written. The author has tried to avoid the use of formulas, graphs, confusing tables, obscure footnotes, and any other material which cannot be clearly understood. Particular attention has been given to the approved repairs as required by the Civil Aeronautics Administration.

This book should furnish all the theory necessary to qualify a person for a Civil Aeronautics Administration aircraft mechanic's certificate.

PREFACE

It should meet text requirements for formal classes studying aircraft maintenance and service. It should also be of great value to members of maintenance crews and to certificated mechanics. In preparing this text the author has kept in mind the needs of students in College Vocational work, Trade Schools, Junior Colleges, High Schools, Aviation Ground Schools, and those in Rehabilitation work.

The author wishes to express his grateful appreciation to the following who have so kindly furnished material which has been of assistance in the preparation of this book: the U. S. Office of Education; Civil Aeronautics Administration; Departments of Education of several states, particularly New York, Pennsylvania, Utah, and Virginia; Douglas Aircraft Company, Inc.; Piper Aircraft Corporation; Stinson Division of Consolidated Vultee Aircraft Corporation; Engineering and Research Corporation; Curtiss-Wright Corporation; Taylorcraft Aviation Corporation; Forest Products Laboratory; Timm Aircraft Corporation; Gladen Products, Division Los Angeles Turf Club; Bell Aircraft Corporation; American Forest Products Industries; Fairchild Engine and Airplane Company; Reynolds Metals Company, Inc.; Aluminum Company of America; The Glenn L. Martin Company; Universal Molded Products Company; The Linde Air Products Company; Cleveland Twist Drill Company; Dow Chemical Company; Allegheny Ludlum Steel Corporation; and Cleveland Pneumatic Tool Company. The author is particularly grateful to those of the above who read and edited parts of the manuscript.

The author wishes to thank his many friends who have so generously contributed their advice and assistance. He also wishes particularly to express his gratitude to Earle R. Hough for his excellent work in the preparation of many of the drawings in this book, and to Mildred Pickrel and Alma Franklin for their untiring patience and cooperation in the preparation of the manuscript.

Frontispiece shows the installation of an engine on a large airplane and is reproduced by courtesy of Douglas Aircraft Company, Incorporated.

R. H. D.

TABLE OF CONTENTS

I	Introduction	1
II	Component Parts of the Airplane	7
III	Glossary	15
IV	Inspection	28
V	General Maintenance of the Airplane	60
VI	The Light Airplane	78
VII	Technical Drawings	109
VIII	General Safety Practices	114
IX	Glues, Glued Joints, and Plywood	118
X	Dopes and Finishes	129
XI	Approved Repairs	135
XII	Woods Used in Aircraft	143
XIII	Metals Used in Aircraft	170
XIV	Alloys Used in Aircraft	188
XV	Fabric-Covered Construction	197
XVI	Plywood Construction	221
XVII	Typical Wood and Fabric Repairs	242
XVIII	Welded Repairs	265
XIX	Drilling, Burring, Filing, and Riveting in Sheet-Metal Repairs	295
XX	Layout and Bend Allowance	313
XXI	Forming Sheet Metal	326
XXII	Assembly	337
Index		349

1 | INTRODUCTION

When a finger is cut or an injury appears on the surface of the body, or an ache or pain makes its appearance, immediate steps are usually taken to cure the ill. The body may go on and perform its daily routine regardless of a cut finger, a minor broken bone, a sore throat, or a stomach ache, but it will not perform as efficiently and effectively as if all parts were healthy and in good condition.

Fig. 1. A light composite airplane being serviced. (Courtesy Piper Aircraft Corporation)

When a part of the body is damaged to any considerable extent, a skilled physician or surgeon is called to give it the necessary care. Neglected small injuries or small illnesses may lead to severe complications, such as infections or serious illnesses, or may lead to complete failure of the body functions.

An airplane and an airplane engine are complicated mechanisms and subject to many small "illnesses" which must be checked before they become serious. If the airplane were a ground mechanism as is an automobile or a carriage, these small illnesses might not be very serious and the airplane might be operated without any particular danger to its occupants.

An automobile may be driven with a tire which is in danger of blowing out so long as the automobile is operated at low speed and driven with care. A fender or hood or a part of the body may be ready to fall off and, if it does, no serious damage results to either the vehicle

Fig. 2. Improved radio equipment, dual engine mufflers and cabin soundproofing make possible the cabin dome speaker which replaces headphones. (Courtesy Consolidated Vultee Aircraft Corporation, Stinson Division)

or its occupants. If the tire goes flat, the engine fails, a fender drops off, a bolt falls out, or a gear is stripped, it is a simple matter to pull the vehicle off to one side of the road and wait for repairs. If the repair cannot be made at the point where the vehicle is, a tow car hauls it to the nearest garage where repairs may be made.

No pilot, however reckless, would think of starting on even a short flight, if he thought there was the slightest danger that any part of the aircraft or engine might fail. The failure of a tire in landing may cause a serious accident. Any failure of a vital part could cause the

INTRODUCTION

entire destruction of the plane and serious injury or death to the occupants.

It is difficult to draw a distinct line between service, maintenance, overhaul, and repair. Service may consist of such simple operations as washing the surface of the airplane, filling the fuel tank, pumping up

Fig. 3. Power plant installation and nose wheel of a light airplane showing cowling removed. (Courtesy Engineering and Research Corporation)

the tires, or renewing the oil supply. The washing of the exterior parts of the airplane, which is a service, may also lead to the preservation of the finish and so fall into the field of maintenance. Maintenance may be thought of as keeping the airplane in such a condition that it is always as good as it was when new. This is not entirely possible as there is always normal wear and deterioration of the various parts of the structure.

AIRCRAFT MAINTENANCE AND SERVICE

All persons responsible for the care and maintenance of the airplane should, insofar as is possible, make certain that the airplane is in such mechanical condition that it will accomplish its mission safely.

Maintenance may be such operations as renewing the finish, replacing small parts which may show signs of wear or failure, making minor repairs, or any other operation necessary to keep the airplane in safe flying condition.

Overhaul does not necessarily include repairs. Overhaul may simply include the disassembly of the structure to determine whether or not

Fig. 4. A modern cargo airplane built almost entirely of wood. The gross weight, when loaded, is more than 36,000 lbs. (Courtesy Curtiss-Wright Corporation)

all of the parts are in such condition that it is safe to operate the aircraft.

Repair may consist of such simple operations as patching the fabric, renewing a cross-member in a wood rib, replacing a few rivets, or removing a scratch from a propeller blade. Repairs, however, may be of the major type such as splicing a spar, welding a new part into a metal fuselage, replacing large parts of fabric or metal skin, or rebuilding a complete major component.

Service may consist of a complete inspection of all parts of the aircraft and engine to determine whether or not they are in safe operating condition.

In general, maintenance and service should include necessary in-

INTRODUCTION

spections, minor repairs, and such other operations necessary to keep the airplane in safe flying condition, but they do *not* include major repairs or complete overhaul. Maintenance and service cannot be clearly defined as to whether the operation may be performed by a person other than a certificated mechanic. Certain repairs which in themselves appear simple and of a more or less minor nature may be such that the services of a certificated mechanic are required. Other services may be performed by persons under the supervision of a certificated mechanic. It is, however, of the greatest importance that every

Fig. 5. A light all metal airplane showing long ailerons and twin tail and cantilever wings. (Courtesy Engineering and Research Corporation)

item of service and maintenance, repair, and overhaul be performed by a person thoroughly familiar with and well trained in the operation he is to perform. It is so necessary that the airplane be maintained in a safe flying condition that Civil Aeronautics Administration experts in aviation have set up complete rules, regulations, and techniques covering the inspection, maintenance, and repair of aircraft and aircraft engines.

Before flight, the Civil Aeronautics Administration requires that an airplane be thoroughly checked by a qualified airman. This preflight check includes all inspections which may be made visually, and covers all parts of the airplane from the fabric covering of the wings to the

AIRCRAFT MAINTENANCE AND SERVICE

water trap in the fuel system. Certain inspections must be made at regular intervals by a certificated aircraft and engine mechanic. It is the responsibility of every pilot, mechanic, and owner to be sure that the complete aircraft meets all of the requirements for safe operation before flight.

It is impossible for any one text to cover every type of aircraft completely. An attempt has been made in this text to give the fundamentals of maintenance and service as required by the Civil Aeronautics Administration for civil aircraft. Much depends upon the knowledge, skill, and conscientiousness of the person charged with maintenance, inspection, and service. It is not only important to know how to perform the operations necessary, but it is just as important to be able to tell when they should be performed and when they have been performed properly.

While many operations pertaining to maintenance and repair have been explained in detail in this text, it is largely assumed that the fundamental techniques of maintenance, repair, and service are under-

Fig. 6. A light all metal airplane having a twin tail and tricycle landing gear. (Courtesy Engineering and Research Corporation)

stood by the mechanic. Most operations of this kind require the services of a certificated aircraft and engine mechanic. While rather complete descriptions are given for the performance of repair operations to wood, sheet metal, and welded structures of aircraft, it is assumed that the mechanic has the fundamental skills and experience to perform them properly.

II COMPONENT PARTS OF THE AIRPLANE

It is not possible in a text of this size to describe every type of aircraft and give the exact method for the maintenance and repair of all of their many parts. Aircraft may be of many types, from a blimp to a jet-propelled airplane. The word, aircraft, when used in this text, will refer to the conventional airplane. Airplanes may be built almost entirely of wood or of metal. Composite aircraft are a combination of wood, fabric, and metal. Light aircraft are usually of the composite type, although some light aircraft are made almost entirely of metal or wood.

The conventional airplane is divided into certain main component

Fig. 7. A modern light airplane of composite construction. Note non-retractable landing gear. (Courtesy Piper Aircraft Corporation)

AIRCRAFT MAINTENANCE AND SERVICE

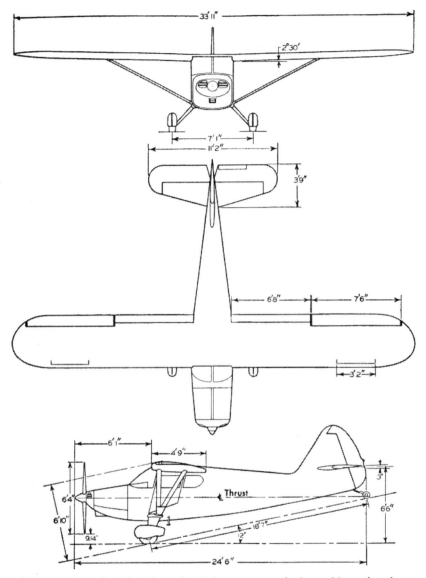

Fig. 8. A three-view drawing of a light post-war airplane. Note the shape of the rudder, ailerons and exceptionally large tail surface. (Courtesy Consolidated Vultee Aircraft Corporation, Stinson Division)

parts. These are the fuselage, which is the main body of the airplane; the wings, which are the supporting surfaces when in flight; the empennage, or tail group; the landing gear; the cowling and engine mount; the controls and control surfaces; and the power plant and propeller.

Within the aircraft are a number of minor assemblies such as the

COMPONENT PARTS OF THE AIRPLANE

ignition system, lubrication systems, electrical systems, instruments, and propellers. The parts of the airplane will be taken up and treated thoroughly enough to enable the mechanic to perform the proper maintenance and service upon the conventional airplane.

Fuselage. The fuselage, which consists of the main body of the airplane, may vary widely in its construction, size, and arrangement, depending upon the use for which it is designed. In the light composite aircraft, the fuselage consists usually of a metal framework of welded steel tubing covered with fabric. This type of fuselage is usually designed to accommodate the pilot and one, two, or three passengers, as well as a limited amount of baggage. Installed in the cockpit, which is a part of the fuselage, are the various controls and instruments.

On large transport aircraft, the fuselage is designed to carry passengers or freight and the pilot's compartment is usually separate from

Fig. 9. A light aircraft constructed almost entirely of metal. Note tricycle landing gear and double tail. (Courtesy Engineering and Research Corporation)

the main cabin for the use of the pilot, copilot, and possibly a navigator and engineer. The fuselage on these large aircraft may be constructed almost entirely of metal or almost entirely of wood.

Wings. The wings of all conventional aircraft are somewhat similar in general design. They are more or less flat on the lower surface and have a cambered upper surface. The wings may be more or less rectangular in shape or may have a decided taper from root to tip. They may be of cantilever construction or have external supports of various kinds.

Fig. 10. Assembly line construction of a light all metal airplane. (Courtesy Engineering and Research Corporation)

Fig. 11. Composite wing construction, wood spars, metal ribs and plywood leading edge. This wing will be covered with fabric. (Courtesy Piper Aircraft Corporation)

COMPONENT PARTS OF THE AIRPLANE

Most wings have two supporting spars running lengthwise of the wing and a number of ribs extending from the leading edge to the trailing edge. The ribs give form to the wing and support to the wing covering which may be of fabric, plywood, or thin sheets of metal.

The internal construction varies widely from the simple light wing which has a front and rear spar and supporting ribs, to the large, all-metal cantilever wings which seem to be almost completely filled with a latticework of metal.

Empennage or Tail Group. Tail groups consist of a vertical fin (sometimes called a vertical stabilizer), a horizontal stabilizer, the rudder, and the elevators. The construction of a tail group varies as does the construction of the other components. The parts of the empennage may be built up of a framework of metal with a fabric or metal covering or a framework of wood with a fabric or plywood covering. Some airplanes may be equipped with more than one tail group. Small control surfaces called trimming tabs may be included in the tail group.

Landing Gear. The landing gear usually consists of the main landing wheels and the tail skid or tail wheel. Or it may consist of the main landing wheels and a nose wheel, as in the typical, tricycle landing gear. The landing gear may be a fixed structure, as is the case on most light airplanes, or the entire landing gear may be retractible. In the retractible type, the wheels and the landing-gear structure may fold into a proper recess in the fuselage or wing, being completely covered by a flap arrangement or extended somewhat into the air stream.

Cowling and Engine Mount. An engine mount consists of a strong, rigid structure by which the engine is attached to the airplane. This structure may be enclosed in part of the fuselage or may be separate from the fuselage and enclosed in a separate streamlined housing called a nacelle. The removable part of the engine and engine-mount covering is called the cowling. The cowling may be equipped with cowl flaps which may be opened and closed to assist in the cooling of the engine.

Controls and Control Surfaces. The control surfaces consist of the ailerons, rudder, elevators, and any attachment to the control surfaces in the way of trimming tabs. The control surfaces vary in design and construction. Many all-metal planes have the control surfaces covered with fabric instead of thin sheet metal. The controls consist of the stick or wheel, rudder bar, and cables connected with the control sur-

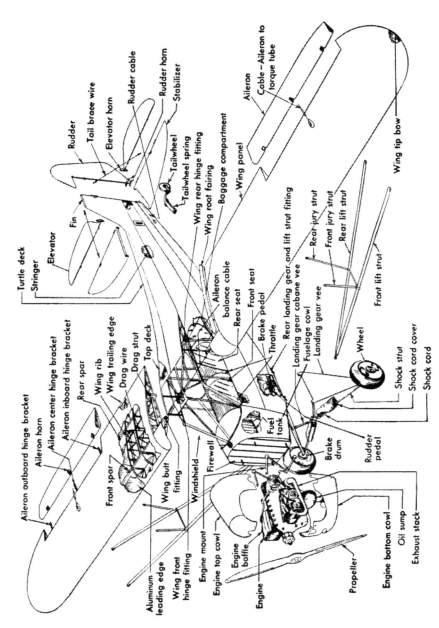

Fig. 12. Visual view of modern light airplane. (Courtesy Piper Aircraft Corporation)

COMPONENT PARTS OF THE AIRPLANE

faces. Flaps and spoilers of various kinds may also be classified as control surfaces. A spoiler is usually a small, adjustable plate arranged to project above the upper surface of the wing to disturb the smooth air flow over the wing. Its action results in the loss of lift and increased drag.

Power Plant. The power plant consists of the aircraft engine and its accessories. The propeller and its accessories are also usually included

Fig. 13. A typical nose wheel and power plant installation on a light airplane. Cowling removed. (Courtesy Engineering and Research Corporation)

as a part of the power plant. Aircraft engines vary in size and type as widely as any other part of the aircraft. The types vary from the light, 4-cylinder, horizontally opposed engines that develop less than 100 hp., to the 4-row radials, developing almost 4000 hp., and the new jet engines which may develop more power than any other aircraft engine.

Conventional engines may be inverted in-line, vertical in-line, inverted V's, upright V's, or of the W type of cylinder arrangement.

The power plant may be broken down into (1) the power unit, which consists of the cylinders, crankcase, crankshaft, and piston and connecting-rod assemblies; (2) the induction system, which consists of the intake manifolds, the carburetion system, the supercharger, and accessories; (3) the accessory group, which consists of the fuel pumps, instrument drive, vacuum pumps, and oil pumps; and (4) the ignition system.

Other parts of the airplane fall into groups and may be called sub-assemblies or subunits. One group of this type is the instrument group which is divided into aircraft instruments and engine instruments. The aircraft instruments may be divided into three groups: (1) the aircraft instruments proper; (2) the flight instruments; and (3) the navigation instruments. Other groups consist of fuel systems, lubrication systems, and electrical systems. The electrical system is broken down into the aircraft electrical system and the engine electrical system or ignition system.

‖‖ GLOSSARY

acetone. A liquid which has the property of absorbing twenty-five times its volume of acetylene under normal temperature and pressure.

acetylene. The fuel used in oxyacetylene welding. It is a compound of hydrogen and carbon, with a flame speed of 330 ft. per sec.

aileron. A hinged or pivoted, movable, auxiliary surface of an airplane, usually part of the trailing edge of the wing, the purpose of which is to maintain the lateral balance of the airplane.

aircraft. Any weight-carrying device or structure designed to be supported by the air, either by buoyance or dynamic action.

airfoil. Any surface, flat or curved, rigid or movable, such as a wing, aileron, or rudder, which catches the air in such a manner as to affect the motion of the plane.

airplane. A mechanically-driven aircraft, heavier than air, fitted with fixed wings and supported by the dynamic action of the air.

air scoop. A scoop or hood designed to catch the air and maintain the air pressure in ballonets, internal-combustion engines, ventilators, etc.

air screw (propeller). A surface so shaped that its rotation about an axis produces a force (thrust) in the forward direction of the axis.

alclad. Sheet Dural with a thin coat of pure aluminum on both sides (aluminum-clad Dural). The purpose of the aluminum is to protect the Dural and prevent corrosion.

alignment. The state of being in line, or to draw into line.

alloy. A mixture or solid solution of two or more metals.

aluminum. Aluminum is one of the 96 chemical elements. It is the principal ingredient in all of the aluminum alloys. A coating of pure aluminum is applied to high-strength aluminum sheets to produce Alclad. The chemical symbol for aluminum is Al. It has a density of 0.0943 lb. per cu. in. The melting point of aluminum is 1220° F.

aluminum alloys. Alloys formed by combining commercially pure aluminum with other metals, namely, copper, silicon, manganese, magnesium, chromium, iron, zinc, and nickel.

angle, attaching. A piece of material bent or formed into an angle and used as a structure for attaching.

angle of incidence. (*See* **angle of wing setting.**) In British terminology, the angle of incidence is equivalent to the American term, "angle of attack."

angle of wing setting. The acute angle between the plane of the wing chord and the longitudinal axis of the airplane. The angle is positive when the leading edge is higher than the trailing edge.

annealing. Heating of an alloy in the solid state to a temperature above the critical range or the solid solution temperature, and allowing it to cool slowly through the critical range. The term, annealing, is used rather broadly sometimes, but in general is to (1) remove internal stresses, (2) induce softness, ductility, and toughness, (3) refine the crystalline structure.

anodize. The artificial oxidation of the surface of aluminum or magnesium alloys by means of an electrochemical treatment.

antidrag wire. A wire designed primarily to resist forces acting parallel to the chord of the wing of an airplane and in the same direction of flight.

assembly. A group of parts fastened together to form a unit which may be installed in another assembly or in the plane itself. The assembly drawing shows the dimensions needed to locate parts with respect to each other and gives instructions for drilling, reaming, trimming, etc.

backhand welding. A method of doing a weld in which the filler rod follows the flame.

backing strip. The material (metal, asbestos, carbon, etc.) used to back up the root of the weld.

base metal (parent metal). The material composing the pieces to be joined by welding.

bevel. An angular surface across the edge of a piece of material.

bevel, closed. The condition of any construction making an angle of less than 90° with 90° perpendicular. Thus, a closed bevel of 10° would specify that the construction formed an angle of 80° with the line perpendicular to the 90° line.

bevel, opened. Same as above except that the angle is the specified number of degrees greater than 90.

blowpipe. The unit of the welding apparatus which brings the two gases together in the correct proportions and thoroughly mixes them before they are released to form the flame.

bonding. A piece of material used to establish an electrical connection between adjoining structures or units.

brashness. A condition of wood characterized by an abrupt failure across the grain without splintering.

brazing. Uniting of metal parts with a copper alloy which has a lower melting point than the metal being joined.

brittleness. That property of a metal which causes it to fracture when deformed.

bucking. The act of holding the bucking bar on the rivet shank to upset the rivet.

GLOSSARY

bulb angle. A right angle of metal having the outer edge of one flange enlarged in a circular manner to lend additional stiffness to the part.

bulkhead. Lateral partitions of an enclosed fuselage.

burring. Removing the ragged edge around holes after drilling.

butt weld. Operation by which the edges of two surfaces are brought together, edge to edge, and are welded along the seam thus formed.

cadmium. One of the chemical elements. A silvery-white crystalline metal whose symbol is Cd. It is commonly used as a plating material (cadmium plate) because of its corrosion-resistant characteristics.

camber. The convex curvature of the surfaces of an airfoil.

cantilever. A kind of wing design so arranged that the entire structural strength of a wing is within the wing itself, requiring no outside bracing.

carburizing. A method of case hardening.

carburizing flame. A gas flame which has the property of introducing carbon into the metal heated.

case hardening. The hardening of the outside of a piece of metal, leaving the inside soft and tough.

center section. The central panel of the wing structures, usually located directly over the fuselage to which the wings are attached on each side.

chamfer. To bevel an edge part way down the side.

chord. A straight line between the leading and trailing edges of an airfoil.

chromodizing. Adding a thin film of aluminum oxide on the surface of Alclad material, by immersion in a chromic acid solution.

cloth. Fabric delivered by the bleachery or finisher before it has been proofed, doped, or specially treated for aeronautic use.

cockpit. An open space in an airplane for the accommodation of pilots or passengers. When completely closed, it is usually called a cabin.

compreg. Plywood, the veneers of which have been treated under pressure and temperature with bonding resin or phenolic resin.

compression failure. Deformations or buckling of the wood fibers resulting from severe stress in compression along the grain.

compression member. A longitudinal strut between spars opposing the pull of wires in the internal drag truss of a wing.

compression rib. A rib which is built particularly strongly with either heavier cap strips or a solid web, or both, for the purpose of withstanding greater compression stress.

concentric. Circles or spheres which have a common center.

conductivity. The property by which heat travels through a metal.

conduit. A tube or trough for receiving and protecting electric wires or cables.

control surfaces. Movable airfoils by means of which the airplane is controlled in flight. They are hinged, auxiliary surfaces and consist of ailerons, elevators, rudders, tabs, and flaps. The control mechanism is the means by which the control surfaces are actuated.

copper. One of the chemical elements. Its symbol is Cu. Copper is used extensively for electrical parts, being a good conductor of heat and electricity. It is used as a base metal for alloying with zinc to form brass and for alloying with tin to form bronze. Copper is quite ductile, malleable, and tenacious.

countersink. To remove material about the end of a hole in an object so that the heads of rivets, bolts, and screws may be flush with the surface of the object.

cowling. A removable covering.

cutting blowpipe. A device used in gas cutting for controlling the gases used for preheating and the oxygen for severing the metal.

cutting tip. A gas blowpipe tip especially adapted for cutting.

cylinder (bottle). A portable container used for the storage of a compressed gas.

datum line. A base line or reference line from which calculations or measurements are taken.

density. The weight of a substance per unit volume.

deposited metal. Metal that has been added by a welding process.

die casting. A method by which molten metal is forced into suitable permanent molds.

dies. Tools, the purpose of which is to impart any desired shape or impress any desired form or design on metals or materials. On those dies which shear a part out of the stock sheet, or form it from the flat sheet, the die is the female portion which the punch enters to perform the required work.

dimpling. Press countersinking with dies for flush riveting in thin material or skin under 0.050 in. thick.

distortion. The deformation of metal due to uneven heating.

dope. The finish applied to aircraft fabric to shrink, preserve, and cause airtightness. Dopes are usually cellulose or collodion compounds, soluble in ether or acetone.

dowel. A small wooden pin used to fasten two pieces of wood in a joint.

drag wire. A wire designed to resist forces acting parallel to the chord of the wing and in the opposite direction of flight.

drill. An instrument for boring holes in hard substances.

durability. A general term for permanence or lastingness.

dural. The alloy of aluminum used most extensively for aircraft production. It is made by the addition of certain small percentages of copper, manganese, magnesium, and silicon —one or all of these being used to form the alloy. In all cases, the aluminum remains above 90 per cent.

edge distance. The distance from the center of a rivet to the edge of sheet or stock.

elasticity. The property of a substance to recover its original size and shape after being deformed.

GLOSSARY

elastic limit. The maximum load a metal will maintain before it begins to permanently deform.

electrolytic. Pertaining to electrolysis, a chemical decomposition caused by an electric current.

elongation. The amount a metal will stretch, when pulled, before it is pulled apart.

empennage. A collective name for the tail-surface group (rudder, elevators, and stabilizers).

exhaust collector ring. A circular duct into which the exhaust gases from the cylinders of a radial engine are discharged.

expansion coefficient. The amount a metal will increase in size per degree rise in temperature.

extrusion. The process or method by means of which the more plastic metals, such as lead, tin, aluminum, zinc, copper, etc., are shaped into intricate cross-sectional shapes by being forced through a die opening of the proper shape.

fabric. The cloth (usually a high-grade cotton) used for covering an airplane.

fairing. A member or structure for the purpose of producing a smooth outline and reducing wind resistance.

fatigue. The loss of strength in a metal due to repeated bending or application of loads.

ferrous. Pertaining to, or derived from, iron.

ferrule. Metal fittings or wire wrappings designed to prevent splitting of wooden parts or loosening of wire terminals, etc.

filler or welding rod. Filler metal, in wire or rod form, used in the gas welding process to add metal to the weld.

fillet weld. A weld made in a corner, as in a lap joint.

filter lens. A colored glass used in goggles, helmets, and shields to exclude harmful light rays.

fin. A fixed vertical surface attached to a part of the aircraft in order to secure stability; for example, a tail fin, skid fin, etc. Fins are sometimes adjustable.

final assembly. This is the assembling of component sections to make up the complete airplane.

firewall. A fire-resistant transverse bulkhead, so set as to isolate the engine compartment from the other parts of the structure and thus confine a fire to the engine compartments as much as possible.

fitting. A general term referring to any one of the many small, and not otherwise named, parts used in fabricating the structure of an aircraft. The most important fittings are those which join two or more major assemblies, such as wing to fuselage. Most fittings are of strong alloys.

flange. Any web stiffening portion of I-beam sections, channel sections, cap strips on wing ribs or spars, etc.

flap. A hinged or pivoted airfoil forming the rear portion of an airfoil. It is used to vary the effective camber.

flux. A chemical material used to dissolve oxides, clean metal, and prevent oxidation during welding.

forging. Metallic parts fabricated by placing a preheated bar or rod under the repeated blows of a hammer and forcing the material to conform to the contours of the steel forming die.

former or false rib. An incomplete rib extending from the leading edge to the front spar to assist in maintaining the front curvature of the wing.

framework. The frame of an airplane exclusive of the skin.

fuselage. The structure, approximately streamlined in form, to which are attached the wings and tail units of an airplane.

gauge. Any suitably shaped standard tool that has been accurately finished to some standard dimension which is used to check or measure the finished dimension of many parts or tools and to measure accurately the distances between various adjacent parts.

glue, *casein.* Glue made from the casein which is precipitated from sour-milk curds by the use of hydrochloric acid.

    cold-setting resin. An adhesive made from synthetic urea and formaldehyde and containing a catalyst or hardener which affects the setting of the glue at temperatures of 21° C. (70° F.) or above.

    hot-setting phenolic type. An adhesive made from phenol or phenolic compounds and formaldehyde, and usually containing a hardener or catalyst. Temperatures of 116° to 160° C. (240° to 320° F.) are usually required to cure these glues.

    water- and mold-resistant casein. An adhesive made from casein, lime, and some alkali-producing sodium salt and containing sufficient quantities of a preservative, such as some sodium chlorophenate, to impart mold resistance to the glue joints.

grain. The direction, size, arrangement, appearance, or quality of the fibers in wood.

    close-grained wood. Wood with narrow and inconspicuous annual rings.

    coarse-grained wood. Wood with wide and conspicuous annual rings.

    cross grain. Grain not parallel with the axis of a piece.

    diagonal grain. Annual rings at an angle with the axis of a piece as a result of sawing at an angle with the bark of the tree.

    edge grain. Edge-grain lumber has been sawed parallel with the pith of the log and approximately at right angles to the growth rings.

    flat grain. Flat-grain lumber has been sawed parallel with the pith of the log and approximately tangent to the growth rings.

    interlocked-grain wood. Wood in which the fibers are inclined in one direction in a number of rings of annual growth, then gradually reverse and are inclined in an opposite direction.

    open-grained wood. Common classification of painters for woods with large pores, also known as "coarse textured."

    plain-sawed. Another term for flat grain.

GLOSSARY

quarter-sawed. Another term for edge grain.

vertical grain. Another term for edge grain.

wavy-grained wood. Wood in which the fibers collectively take the form of waves or undulations.

gusset. A small connection plate, usually of thin plywood, placed over a joint to give it strength.

hardness. The ability of a metal to resist permanent deformation.

hardness tests for metal. Brinell Hardness Test, Rockwell Hardness Test, and Shore Scleroscope Hardness Test. There are other hardness tests, but the three listed are the ones in most common use.

hardwoods. The botanical group of trees that are broad leaved. The term has no reference to the actual hardness of the wood.

heart, heartwood. The wood extending from the pith to the sapwood.

heat treatment. Changing the physical properties of metals and alloys by heating and cooling under various conditions.

heat treatment of aluminum alloys. A very accurately controlled process of heating, quenching, and aging of the alloys, which develops the maximum hardness and strength characteristics of the hard aluminum alloys.

honeycomb. Checks, often not visible at the surface, that occur in the interior of a piece.

horn. A short lever attached to a control surface of an aircraft to which the operating wire or rod is connected.

hot short. The condition in which metals become extremely weak at temperatures below the melting point.

inconel. A registered trade mark of the International Nickel Company, Inc. The name, Inconel, is applied to a nickel chromium iron alloy. It contains approximately 80 per cent nickel, 14 per cent chromium, and 6 per cent iron. It has physical properties similar to stainless steel. It cannot be hardened by heat treatment, but the effects of cold working (strain hardening) can be relieved by annealing.

inspection hole. A small hole having an easily removable cover.

jig. A pattern, form, or framework, accurately dimensioned and aligned, in which identical structures of parts can be produced to meet a standard.

joint. A union of two pieces of stock.

joint splice. A joint which consists of two pieces of wood placed end to end and covered with pieces of wood or fishplates bolted or nailed to each side.

kerf. The slot made by the cut of a saw.

kiln. A heated chamber for drying lumber.

knot. That portion of a branch or limb that has become incorporated in the body of a tree.

laminated wood. An assembly built up of plies of laminations of wood that have been joined either with glue or with mechanical fastenings. Distinguished from plywood by the fact that the grain of the wood is in the same direction in all plies.

AIRCRAFT MAINTENANCE AND SERVICE

leading edge. The foremost edge of an airfoil or propeller blade, also called "entering edge."

lightening hole. A hole cut in a solid rib or bulkhead to reduce weight.

longeron. A longitudinal member of the framework of an airplane fuselage or nacelle.

loom. The web sections of wing ribs.

lucite. Trade name for a transparent plastic material which softens when heated. Used for windows, nose, etc., on airplanes, and for tooling purposes.

lumber. The product of the saw and planing mill not further manufactured than by sawing, resawing, and passing lengthwise through a standard planing machine, crosscut to length and matched.
  factory and shop lumber. Lumber intended to be cut up for use in further manufacture. It is graded on the basis of the percentage of the area which will produce a limited number of cuttings of a specified or a given minimum size and quality.
  yard lumber. Lumber that is less than 5 in. in thickness and is intended for general building purposes.
    boards. Yard lumber less than 2 in. thick, 8 in. or more in width.
    dimension. All yard lumber except boards, strips, and timbers; that is, yard lumber 2 in. and less than 5 in. thick, and of any width.
    timber is more than 5 in. thick and more than 5 in. wide.
    strips. Yard lumber less than 2 in. thick and less than 8 in. wide.

magnetic inspection. The determination of flaws at or near the surface of a ferromagnetic metal by magnetizing the metal or member to be tested and then dusting dry, fine iron particles onto the surface.

main beam. A rigid body designed to transmit transverse loads in shear and/or bending to its points of support.

manifold. A pipe with outlets or branches to which several cylinders of gas may be connected to supply gas to a number of blowpipes.

masonite. Trade name for an alkaline plastic made from wood pulp.

melting point. The temperature at which a substance changes from the solid to the liquid state.

member. Any essential part of a structure or machine.

metal. Any of the group of elements which are typically fusible and opaque, are good conductors of electricity, and have a typical metallic luster.

moisture content of wood. Weight of the water contained in the wood, expressed in a percentage of the weight of the oven-dried wood.

mold. The cavity in which anything is shaped. A frame or body on or about which something is made.

mold line. A line formed by the intersecting of two planes (flat surfaces). In the case of an angle where the bend is a radius, the mold line will be in space at the point where

GLOSSARY

the outside surfaces of the legs of the angle would meet if they were extended.

monel metal. An alloy of nickel, 67 per cent; copper, 28 per cent; iron, manganese, silicon, and other metals, 5 per cent. Monel metal is extremely resistant to corrosion.

monocoque fuselage. A type of fuselage construction wherein the structure consists of a thin shell of wood, metal, or other material supported by ribs, frames, belt frames, or bulkheads, but usually without longitudinal members other than the shell itself. The structure is built so that the shell carries the greater part of the stresses to which the fuselage is subjected.

motor mount. A support on which to mount the motor of an airplane.

nacelle. An enclosed shelter for passengers or for an engine. A nacelle is usually shorter than a fuselage and does not carry the tail unit.

neutral flame. The neutral flame is created by burning acetylene with oxygen in such proportions that all particles of carbon and hydrogen in the fuel gas are consumed. It produces a temperature of approximately 6300° F.

nose. The bow of an airplane, usually the front of the fuselage or hull. It also refers to the front end of a wing rib or the leading edge of a wing.

oilcan. (Colloquial.) In regard to the metal skin of an airplane. An expression meaning that the skin is slightly bulged up between rows of rivets on a metal skin structure. This bulge can often be pushed back and forth by applying a slight pressure to the metal and the so-called "oilcan" will react the same as the bottom of an oil can. Such a condition is due to the metal being forced out of line by a row of rivets and must be avoided.

oxide. The coating or scale formed on metal when combined with oxygen.

oxidizing flame. Contains an excess of oxygen which combines with the metal. This action is damaging to many metals.

oxy-acetylene welding. A gas welding process wherein the welding heat is obtained from the combustion of oxygen and acetylene.

oxygen. An odorless, colorless, tasteless gas, forming approximately 21 per cent of the atmosphere.

peening. The mechanical working of metal by means of hammer blows.

penetration. The depth of fusion obtained in a welded joint.

pickling. The removal of surface contamination from metals or alloys by dipping in a solution.

pitch pocket. An opening extending parallel to the annual rings of growth which usually contains, or has contained, pitch.

plastic. A general term applied to low-density synthetic resins and materials in solid form which can be formed or molded, under suitable conditions, to desired shapes. The chemical ingredients of plastics unite by means of polymerization.

Plexiglas. The trade name for a transparent plastic material which softens when heated. Used for windows, nose, etc., on airplanes; used for tooling purposes.

ply. A thickness or layer of wood or other substance.

plywood. An assembly made of three or more layers of veneer joined with glue.

porosity. The presence of gas pockets or inclusions.

power brake. A sheet-metal brake operated by power.

preheating. Heat applied prior to welding, cutting, or forming operations.

primer. Paint used for a first, or prime, coating. Zinc chromate is usually used as a primer in airplane construction.

ream. To widen the opening of (a hole); to bevel out. To enlarge (a hole) with a reamer.

rib. A light, curved, metal part mounted in a fore-and-aft direction within a surface. The ordinary ribs give the surface its camber, carry the fabric, and transfer the lift from the fabric to the spars.

rib, box. A rib built in the form of a solid box at the junction of the wing and fuselage.

rib, form. The lighter, form-giving parts used in airfoil construction.

rib, ordinary. A light, curved, wooden part mounted in a fore-and-aft direction within a surface. The ordinary ribs give the surface its camber, carry the fabric, and transfer the lift from the fabric to the spars.

rig (airplane). To assemble, adjust, and align the parts of an airplane.

ring, annual-growth. The growth layer of a tree put on in a single growth year.

ring cowling. A ring-shaped cowling placed around a radial air-cooled engine to reduce its drag and improve cooling.

ripple. Tiny wavelike formations in the finished weld caused by successively cooling puddles.

rivet draw. A special bar with a hole in one end large enough to slip over the rivet shank which is used to draw or set the sheets together.

rivet gauge. A tool for inspecting the height and diameter of a rivet shank after upsetting or driving. A flush type of rivet gauge is used for inspecting the height of the manufactured head of the rivet projecting above or below the surface.

riveting. The method of permanently fastening or joining sheets and parts together with rivets.

riveting, blind. Upsetting rivets from one side only.

riveting gun. A compressed-air or pneumatic hammer used for riveting.

riveting, hand. Upsetting rivets by hand hammer or hand squeezer.

rivet set. A tool to drive or upset rivets. It is a tool with a cup-shaped depression in one end that must fit the head of the rivet being used.

GLOSSARY

sap. All the fluid in a tree.

sapwood. The layers of wood next to the bark, usually lighter in color than the heartwood, ½ in. to 3 or more in. wide that are actively involved in the processes of the tree.

seasoning. Removing moisture from green wood in order to improve its serviceability.
air dried or air seasoned. Dried by exposure to the air, usually in a yard, without artificial heat.
kiln dried. Dried in a kiln with the use of artificial heat.

shake. A separation along the grain, the greater part of which occurs between the rings of annual growth.

skip welding. The method of welding a seam where short sections are welded, leaving unwelded sections.

slag. The coating of impurities formed on molten metal consisting largely of oxides and silicates.

softwoods. The botanical group of trees that have needle-like or scale-like leaves and are evergreen for the most part.

solder. A metal or metallic alloy used, when melted, to join metallic surfaces. Solders which melt readily are soft solders; others fusing at a red heat are hard solders.

span. The maximum distance, measured parallel to the lateral axis from tip to tip of an airfoil, of an airplane wing inclusive of aileron, or of a stabilizer inclusive of elevator.

spar, main. The front spar within the wing surface and to which all ribs are attached, such spar being the one situated nearest to the center of pressure.

spar, rear. A spar within the wing and to which the ribs are attached, such spar being situated at the rear of the center of pressure and at a greater distance from it than is the main spar. It transfers less than half of the lift from the ribs to the bracing.

specific gravity. The ratio of the weight of a body to the weight of an equal volume of water at some standard temperature.

spinner. The fairing of approximately conical shape which is fitted over the propeller hub and revolves with the propeller.

splice. The adjoining of the ends of two pieces of rope, wire, or wood in such a manner that the splice (joint) is just as strong as the material spliced.

split. A lengthwise separation of the wood due to the tearing apart of the wood cells.

spot welding. A weld, either gas or electric, which is not a continuous seam weld, but is made only at spots. Spot welding generally refers to electrical resistance welding.

spring wood. The portion of the annual growth ring that is formed during the early part of the season's growth.

stainless steel. The term, stainless steel, refers to a series of alloys containing chromium and nickel. These alloys have special properties such as high resistance to corrosion and strength at high temperatures.

strain. The effect of the application of external forces to a body.

stress. Any internal force acting within a structure.

strut. A compression member of a truss frame; for instance, the vertical members of the wing truss of a biplane (interplane struts) and the shorter vertical and horizontal members separating the longerons in the fuselage.

strut, fuselage. A strut holding the fuselage longerons apart. It should be stated whether top, bottom, or side. If side, then it should be stated whether right or left.

subassembly. This consists of an assembly of component parts to make up a section of an airplane.

summer wood. The portion of the annual growth ring that is formed during the latter part of the yearly growth period. It is usually denser and stronger mechanically than spring wood.

tack weld. A very short weld used to hold parts being formed in place while welding.

tailboom. A spar connecting the tail surfaces and the main supporting surfaces of the airplane.

tap. A tool for forming an internal screw thread. In an electric circuit, a point where a connection may be made.

template. A full-sized pattern from which structural sections are marked out. It may be of paper, cardboard, plywood, or metal.

tensile strength. The resistance to forces tending to pull the material apart.

tip (blowpipe). That part of the blowpipe from which the mixed gases issue to form the flame.

trailing edge. The rearmost edge of an airfoil or of a propeller blade.

undercut. The condition where the base metal has been fused without sufficient metal being added, leaving a groove along the edge of the weld.

upsetting. The operation of setting or driving the rivet shank to the correct height and diameter.

veneer. Thin sheets of wood.
rotary-cut veneer. Veneer cut in a continuous strip by rotating a log against the edge of a knife in a lathe.
sawed veneer. Veneer produced by sawing.
sliced veneer. Veneer that is sliced off by moving a log, bolt, or flitch against a large knife.

wane. Bark, or lack of wood from any cause, on the edge or corner of a piece.

warp. Any variation from a true or plane surface.

web. A structure member used to give form or to transmit a load.

welding hose. The hose lines making connections between the blowpipe and the regulators, conveying the two gases to the blowpipe.

wing. The main lifting and supporting surface of the airplane in flight,

GLOSSARY

designated as right and left. The right and left sides of an airplane are relative to the right and left hand of the pilot seated in the cockpit.

wing rib. A chordwise member of the wing structure of an airplane used to give wing-section form and to transmit the load from the fabric to the spars.

wing tip. The outer or outboard end of a wing.

workability. The degree of ease and smoothness of cut obtainable with hand or machine tools.

yield strength. The point at which permanent deformation in the structure begins to take place.

IV INSPECTION

The most important part of aircraft maintenance and service is inspection. It is sometimes more difficult to inspect an airplane and to decide whether or not it needs maintenance and service than it is actually to perform the maintenance and service operations. It requires a higher degree of knowledge to inspect than it does to repair. Inspection requires the highest type of judgment. Judgment can only be developed by training and experience. The unskilled person may look an airplane over with extreme care and be unable to tell whether or not the airplane is in safe flying condition.

Every mechanic has had the experience of examining some part of a machine and then admitting to himself that he does not know whether

Fig. 14. Maintenance crew at work on a light airplane. (Courtesy Piper Aircraft Corporation)

INSPECTION

or not the part is in good operating condition. An inspector must be able not only to tell that a part has failed, but also to determine whether there is any indication of future failure. He must be able to tell whether or not the damage is severe enough to require replacement of a part, or whether the part may be retained in service after the proper repairs have been made. He must be able to tell with a high degree of accuracy whether or not the amount of wear shown by a part justifies its replacement. It is the inspector's problem to state definitely what shall be done in the way of maintenance and service to keep the airplane in an absolutely safe flying condition.

The airplane covering and all other parts may appear to the layman to be as good as ever, while the inspector may condemn it immediately and require that it be replaced. The inspector's word should

Fig. 15. Riveting the trailing edge of a metal rudder with a squeeze-type riveter. The large tail surface contributes to stability. The vertical fin is 7 feet high. (Courtesy Consolidated Vultee Aircraft Corporation, Stinson Division)

be final, and all repairs and maintenance operations should be performed in accordance with the rules and regulations of the Civil Aeronautics Administration. The Civil Aeronautics Administration gives the following definition of terms.

AIRCRAFT MAINTENANCE AND SERVICE

Maintenance. Maintenance is a term applied to the process necessary for the retention of the airplane structure, power plant, equipment, and accessories in satisfactory condition for flight. The goal of maintenance is to preserve as nearly as possible the entire aircraft and the material from which it is constructed in the condition in which it was immediately after building. The structure should be protected insofar as possible from the effect of the elements, wear due to normal use, and damage as a result of accident or abnormal use. When any part of the aircraft can no longer be maintained in a safe flying condition, that part, or the whole, must be replaced. The airplane is a highly stressed structure, and failure of any part may cause the loss of the whole structure.

The importance of maintenance cannot be overestimated. The primary functions of maintenance are (1) inspection, (2) upkeep, (3) repairs, and (4) overhaul.

Inspection consists of a close examination of the entire structure to determine whether or not each part of it is capable of performing the functions for which it is designed.

Upkeep includes the normal care of the aircraft, such as the cleaning, corrosion prevention, maintenance of alignment, and adjustments of clearance.

Repair includes the routine operations of mending and renewal of a part which can be performed by the certificated aircraft and engine mechanic.

Overhaul includes the complete disassembly, repair or replacement of parts, reassembly, and testing.

It is only by means of inspection that it is possible to determine and prevent future failures. It is only by means of inspection that oversights or carelessness on the part of the maintenance crew may be discovered. All machines, including an aircraft, are subject to normal wear in use, abuses in operation, and the effects of weather.

Operational abuses include all abuses which might seem to fall under operation. Some operational abuses are racing the engine, improper maneuvers in the air, allowing the engine to overheat, rough handling of instruments and accessories, improper starting and stopping of the engine, improper use of electrical or mechanical equipment, undue exposure to heat, cold, moisture, or other weather conditions, and rough handling either on the ground or in the hangar. It is only by careful inspection that such items as a small hole in the fabric, a frayed control

INSPECTION

cable, lack of safety wire, a loosened nut, a worn pin, a nicked propeller, or dozens of other minor things are discovered which, if not corrected, might lead to serious damage or failure.

It has been determined by experience when inspections should be made and of what they should consist. Many parts of the airplane have

Fig. 16. During manufacturing the airplane receives thousands of inspections. (Courtesy Engineering and Research Corporation)

been found to function properly over long periods of time without showing any signs of failure. Other parts must be checked frequently to determine whether or not they are still in good condition. For instance, under normal conditions, there is no need to examine the crankshaft for hundreds of hours of normal service while, on the other hand, the aircraft fabric covering may well be examined before each flight to determine whether or not damage has occurred. Fabric may crack or tear due to being struck by gravel during take-off or landing or develop cracks or tears with no apparent cause other than the normal vibration during flight.

An airplane should be given a daily or preflight inspection. This inspection should be made by a properly qualified person. It is not necessary to inspect an airplane each day if the airplane has not been

AIRCRAFT MAINTENANCE AND SERVICE

Fig. 17. A light airplane showing the all metal fuselage, cowling, wing attachment and landing gear. (Courtesy Engineering and Research Corporation)

or is not to be flown. However, unless the aircraft is in storage, it should be inspected at least once every week to determine that the airplane is being properly maintained.

In making any inspection, an inspection record should be used. It is not sufficient that the inspector simply use his judgment as to what should be examined, but he should examine each and every item shown on the inspection record and record his findings.

Each time an aircraft is flown, a flight record should be filled out upon which the pilot should not only note improper functioning but also should check the proper functioning of the items shown in the flight-record report.

A reproduction of the Civil Aeronautics Administration's Daily Flight Inspection Record and Flight Record Report are shown at the end of this chapter. It is the pilot's responsibility to see that the airplane which he is to fly has been properly inspected, maintained, and serviced. The fact that the airplane has been O.K.'d by the inspector and maintenance crew does not relieve the pilot of this responsibility.

All forms should be carefully filled out and signed by the proper

INSPECTION

person. It is important that both the aircraft and engine log books be carefully maintained. The responsibility for maintaining these books rests with the owner of the aircraft. The statement of the pilot that the aircraft performs satisfactorily is not sufficient to justify lack of regular inspections. Many things may develop which would lead to failure but may not be apparent to the pilot while in flight. The pilot himself should always make a careful inspection of the airplane which he has to fly before each flight. He should determine the proper operation of the engine and all engine instruments. He should test the controls for proper functioning. He should examine carefully the condition of brace wires, fabric, tires, struts, wheels, control surfaces, security of all inspection plates, proper fastening of cowling, and the proper loading of equipment and baggage. He must also see that all loose objects in the airplane are properly secured.

It is a good practice at the end of 25- and 50-hour periods of flight operation to have the entire aircraft inspected by a certificated aircraft and engine mechanic. This inspection should include not only the visual inspection, such as is performed by the pilot, but also the removal of all inspection plates and the opening of inspection apertures to determine the condition of the various parts. The Civil Aeronautics Administration requires a 100-hour check by a certificated aircraft and engine mechanic.

Proper alignment is important if the airplane is to perform properly in flight. It is the duty of the inspector to determine by measurements whether or not the various parts are properly aligned in relation to each other. Cables, tie rods, and brace wires tend to stretch during use, while lock nuts, turnbuckles, nuts, safety wires, and cotter pins tend to work loose. It may be necessary at regular intervals to place the aircraft in flying position on the ground in order to determine that the parts are properly aligned.

As each component part of the aircraft is discussed, the inspection procedure will be given in more complete detail. The inspector should always have available and make use of the Manufacturer's Service and Maintenance Manual for the airplane upon which the inspection is being made. All recommended repairs and adjustments should be made in accordance with the manufacturer's specifications and the requirements of the Civil Aeronautics Administration. In the following discussion of the various parts, it will be impossible to give detailed instructions for every airplane. Insofar as possible, typical procedures

AIRCRAFT MAINTENANCE AND SERVICE

which may be followed for the more common types of aircraft will be given.

Fuselage. The fuselage is the main body of the airplane. In most light aircraft it consists of a framework built of welded steel tubing. Wood may be used for nonstructural braces or fairings. This type of fuselage is usually covered with fabric which is stitched into place and treated

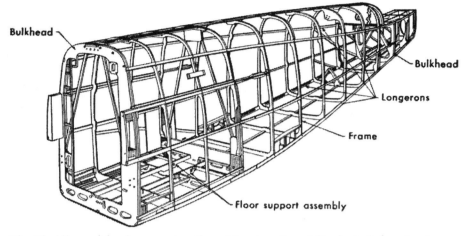

Fig. 18. All wood fuselage construction. (Courtesy Forest Products Laboratory)

with airplane dope to tighten it and to make it airproof and waterproof. The dope is usually covered with a pigmented finish and may be waxed and polished.

In some aircraft, the fuselage may be largely constructed of wood. Wood fuselages are usually of the monocoque type and covered with a stressed skin of plywood. The plywood covering is treated with a suitable finish to protect the plywood from the elements and give it a pleasing appearance.

When constructed of metal, the fuselage is usually of the monocoque type of construction and has a stressed skin of thin metal sheets. The fuselage is the main supporting structure of the aircraft and should be carefully inspected for damage or signs of failure in its main supporting parts. The main longitudinal members of the fuselage are the longerons. The longerons in the welded steel fuselage are braced by means of diagonal and cross braces, and the whole fuselage is divided into bays and trusses. Lengthwise stringers of wood may be installed to give additional shape to the fuselage. When the monocoque type of construction is used, the forming members are the bulkheads which

INSPECTION

Fig. 19. An all metal center section to which the wings are attached. (Courtesy Engineering and Research Corporation)

are usually connected by longitudinal members in the form of longerons or stringers. This type of fuselage does not usually have cross members. The main strength of the structure is in the bulkheads and the skin itself.

For inspection, maintenance, and repair purposes, hulls and floats may be included in the fuselage inspection. The following are suggested inspection procedures.

1. Check all accessible parts of the exterior and the interior of the fuselage, hulls, and floats for the following:
   a. Bent longerons and braces
   b. Cracks in tubing or signs of corrosion
   c. Loose members, bolts, rivets, nuts, safety wires, cotter pins
   d. Proper attachment of inspection plates, apertures, or aircraft components
   e. Condition of protective coatings such as paint, varnish, tape, or chafing preventive devices
   f. Condition of identification markings of control cables, wires, or fittings
   g. Loose bonding or corroded bonding connections.

2. Check for cleanliness throughout; clean and grease exposed metal parts that may come in contact with salt water.

AIRCRAFT MAINTENANCE AND SERVICE

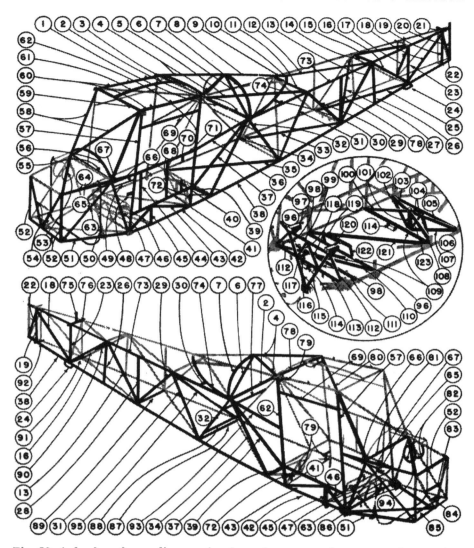

Fig. 20. A fuselage frame diagram showing reference number, name of part, size, material and number needed. (Courtesy Taylorcraft Aviation Corporation)

Fig. 20. Fuselage Frame Diagram. (**Key to Fig. 20**).

REF. NO.	NAME	SIZE (IN INCHES)	MATERIAL	NO. PER ASSEMBLY
1	Tube Assy.-Rear Stab. Cross			1
2	Tube-Sta. 10–12, 10′–12′	½ × .035 × 14¾	AN–T–4	1 ea.
3	Tube-Sta. 9–14, 9′–14′	½ × .035 × 17¾	AN–T–4	2
4	Lug-Fairing Mounting	⅜ × ⅜ × .040 × 1⅜	AN–S–11	2
5	Tube-Sta. 8′–15	½ × .035 × 33½	AN–T–4	1
6	Tube-Sta. 7′–8	½ × .035 × 44	AN–T–4	1

INSPECTION

Fig. 20. Fuselage Frame Diagram. (**Key to Fig. 20**).

REF. NO.	NAME	SIZE (IN INCHES)	MATERIAL	NO. PER ASSEMBLY
7	Tube-Sta. 6–11, 6'–11'	5/8 × .035 × 126 3/8	AN–T–4	2
8	Tube-Sta. 7'–16	5/8 × .035 × 38	AN–T–4	1
9	Tube-Sta. 5–6, 5'–6'	1/2 × .035 × 11 1/2	AN–T–4	2
10	Tube-Sta. 32–6', 32–6	1/2 × .035 × 19	AN–T–4	2
11	Tube-Sta. 4–5, 4'–5'	1/2 × .035 × 23	AN–T–4	2
12	Clip-Fairing Center Section	1/2 × .049 × 1 1/2	AN–S–11	2
13	Tube-Sta. 4'–32, 4–32	1/2 × .035 × 27	AN–T–4	2
14	Tube-Sta. 6–17, 6'–17'	3/4 × .035 × 35 1/4	AN–T–4	2
15	Lug-Floorboard Long	1/2 × .049 × 1 3/16	AN–S–11	2
16	Tube-Sta. 3–4, 3'–4'	3/4 × .035 × 30 1/4	AN–T–4	2
17	Tube-Sta. 3'–4	7/8 × .035 × 40 3/4	AN–T–3	1
18	Support-Window and Door	1/2 × .020 × 30 11/32	Low Carbon Steel Square Tube	2
19	Tube-Sta. 3–3'	1 1/4 × .058 × 28 5/8	AN–T–3	1
20	Tube-Sta. 3–20, 3'–20'	1 × .035 × 48	AN–T–3	2
21	Tube-Sta. 2–3, 2–3'	3/4 × .035 × 34	AN–T–3	2
22	Tube-Sta. 22–23, 22'–23'	5/8 × .035 × 10 3/4	AN–T–4	2
23	Fabric Support-Rear Cowl A Assy.			
24	Ring-Instrument Panel Support	1/4 × 1/2 × .050 × 42	AN–S–11	1
25	Tube-Sta. 1–22, 1'–22'	5/8 × .035 × 20 1/8	AN–T–4	2
26	Tube-Sta. 1–2	5/8 × .035 × 16	AN–T–4	1
27	Tube-Sta. 1–21, 21–21'–1', 1'–1	5/8 × .035 × 22 43/64	AN–T–3	4
28	Tube-Sta. 1–27, 1'–27	5/8 × .035 × 23 3/4	AN–T–3	2
29	Tube-Sta. 20–27, 20'–27	3/4 × .035 × 30	AN–T–4	2
30	Tube-Sta. 20–21, 20'–21'	3/4 × .035 × 25 1/4	AN–T–3	2
31	Tube-Sta. 3–21, 3'–21'	3/4 × .035 × 59	AN–T–3	2
32	Channel-Support Rear Cowl	1/4 × 1/2 × .050 × 40 1/8	AN–S–11	2
33	Tube-Sta. 1'–2	5/8 × .035 × 16	AN–T–4	1
34	Tube-Sta. 20–22, 20'–22'	5/8 × .035 × 24	AN–T–4	2
35	Tube-Sta. 17–20, 17'–20'	7/8 × .035 × 56 1/4	AN–T–3	2
36	Tube-Sta. 20–25, 20'–25'	5/8 × .035 × 13	AN–T–3	1 ea.
37	Tube-Sta. 19–25, 19'–25'	5/8 × .035 × 7 3/8	AN–T–4	2
38	Tube-Sta. 18–25, 18'–25'	3/4 × .035 × 20 5/8	AN–T–4	2
39	Tube-Sta. 24–26, 24'–26'	7/8 × .035 × 32	AN–T–3	2
40	Tube-Sta. 18–26, 18'–26'	7/8 × .035 × 10	AN–T–4	1 ea.
41	Tube-Sta. 17–26, 17'–26'	5/8 × .035 × 13 3/4	AN–T–3	2
42	Tube-Sta. 6–28, 6'–28'	3/4 × .035 × 38 1/4	AN–T–4	2
43	Tube-Sta. 4–26, 4'–26'	7/8 × .035 × 35	AN–T–4	2
44	Tube-Sta. 12–17, 12'–17'	3/4 × .035 × 143 3/8	AN–T–3	2
45	Tube-Sta. 6–26, 6'–26'	7/8 × .035 × 35 1/2	AN–T–3	2
46	Tube-Sta. 6–6'	1/2 × .035 × 28	AN–T–4	1
47	Tube-Sta. 6–16, 6'–16	1/2 × .035 × 30 1/2	AN–T–4	2
48	Support-Longeron Front	3/8 × 3/8 × .040 × 27 1/16	AN–S–11	2
49	Tube-Sta. 7–16, 7'–16'	1/2 × .035 × 30	AN–T–4	2
50	Tube-Sta. 7–15, 7'–15'	1/2 × .035 × 30 3/8	AN–T–4	2
51	Tube-Sta. 8–15, 8'–15'	1/2 × .035 × 30 1/4	AN–T–4	2
52	Tube-Sta. 14–15'	1/2 × .035 × 46 1/2	AN–T–4	1
53	Tube-Sta. 8–14, 8'–14'	1/2 × .035 × 30 1/4	AN–T–4	2
54	Handle-Tail Lift	3/8 × .035 × 8	AN–T–4	2
55	Tube-Rudder Guard Cross	5/8 × .058 × 6 1/4	AN–T–4	2
56	Tube-Sta. 10–14, 10'–14'	1/2 × .035 × 30 1/4	AN–T–4	2
57	Tube-Sta. 10–13, 10'–13'	1/2 × .035 × 14 7/8	AN–T–4	2
58	Support-Longeron	3/8 × 3/8 × .040 × 23 1/8	AN–S–11	2

AIRCRAFT MAINTENANCE AND SERVICE

Fig. 20. Fuselage Frame Diagram. (Key to Fig. 20).

REF. NO.	NAME	SIZE (IN INCHES)	MATERIAL	NO. PER ASSEMBLY
59	Aileron Pulley Bracket Assy.			4
60	Tube-Rear Cowl Channel	$3/8 \times .028 \times 11^{5}/_{16}$	AN-T-4	2
61	Tube-Sta. 2–22, 2–22'	$5/8 \times .035 \times 17^{1}/_{2}$	AN-T-4	1 ea.
62	Bracket Assy.-Front Throttle			
63	Tube-Control Panel Support Horizontal	$1/2 \times 1/2 \times .028 \times 31^{1}/_{16}$	AN-T-4	1
64	Tube-Sta. 4–4'	$1 \times .035 \times 28^{5}/_{8}$	AN-T-3	1
65	Tube-Sta. 5–5'	$1/2 \times .035 \times 32^{1}/_{4}$	AN-T-4	1
66	Tube-Sta. 6'–7	$1/2 \times .035 \times 38^{5}/_{8}$	AN-T-4	1
67	Clevis-Fairing	$1/2 \times .035 \times 2^{1}/_{2}$	AN-S-11	2
68	Tube-Sta. 8'–8	$1/2 \times .035 \times 12^{1}/_{4}$	AN-T-4	1
69	Tube-Sta. 8'–9	$1/2 \times .035 \times 25^{3}/_{8}$	AN-T-4	1
70	Tube-Sta. 9–14'	$1/2 \times .035 \times 18^{3}/_{4}$	AN-T-4	1
71	Post Assy.-Stab. and Fin			1
72	Post-Fin Front	$3/4 \times .035 \times 11^{11}/_{16}$	AN-T-4	1
73	Tube-Tailpost	$3/4 \times .035 \times 17$	AN-T-4	1
74	Mount Assy.-Tail Spring-Front			
75	Tube-Sta. 14–14'	$1/2 \times .035 \times 5^{7}/_{8}$	AN-T-4	1
76	Support-Rudder Cable Pulley Bracket	$1/2 \times .035 \times 9$	AN-S-11	1
77	Tube-Sta. 15–15'	$1/2 \times .035 \times 17^{1}/_{8}$	AN-T-4	1
78	Tube-Sta. 15–16'	$1/2 \times .035 \times 40^{1}/_{2}$	AN-T-4	1
79	Tube-Sta. 17–17'	$7/8 \times .035 \times 27^{3}/_{4}$	AN-T-4	1
80	Tube-Sta. 16–17'	$5/8 \times .035 \times 42^{3}/_{4}$	AN-T-4	1
81	Support-Control Panel Rear Upright	$1/2 \times 1/2 \times .050 \times 14^{11}/_{16}$	AN-S-11	1
82	Support-Control Panel	$1/2 \times 1/2 \times .028 \times 15^{7}/_{8}$	AN-S-11	1
83	Tube-Front Brake and Rudder Support	$5/8 \times .065 \times 23$	AN-T-4	1
84	Support-Battery Lower	$1/2 \times 1/2 \times .050 \times 4$	AN-S-11	1
85	Support-Battery Upper	$1/2 \times 1/2 \times .050 \times 9^{3}/_{16}$	AN-S-11	1
86	Strap-Collector Tank	$1/2 \times .035 \times 22^{27}/_{32}$	AN-S-11	2
87	Tube-Sta. 22–22'	$5/8 \times .035 \times 25$	AN-T-4	1
88	Bracket Assy.-Throttle			1
89	Bracket-Trim Tab Indicator Support	$3/_{16} \times 1^{3}/_{16} \times .035 \times 3$	AN-S-11	1
90	Tube-Sta. 17–17'	$7/8 \times .035 \times 27^{3}/_{4}$	AN-T-4	1
91	Tube-Sta. 7–7'	$1/2 \times .035 \times 22^{5}/_{8}$	AN-T-4	1
92	Tube-Sta. 4–6, 4'–6'	$5/8 \times .035 \times 27^{1}/_{4}$	AN-T-4	2
93	Slide Assy.-Right			1
94	Tube-Rear Seat Truss Top	$3/4 \times .049 \times 18$	AN-T-3	1
95	Tube-Rear Truss Inner Diagonal	$3/8 \times .035 \times 8^{27}/_{32}$	AN-T-4	2
96	Slide Assy.-Left			1
97	Tube-Sta. 18–28, 18'–28	$1/2 \times .035 \times 17^{5}/_{8}$	AN-T-4	2
98	Brake and Rudder Pedal Mounting-Left			1
99	Slide Assy.-Seat Adjustment-Left			1
100	Brace Tube-Shock Truss	$1/2 \times .035 \times 6^{7}/_{8}$	AN-T-4	1 ea.
101	Tube-Sta. 20–30, 20'–30'	$3/4 \times .035 \times 8^{1}/_{2}$	AN-T-3	2
102	Tube-Sta. 29–30, 29'–30'	$1/2 \times .035 \times 6^{5}/_{8}$	AN-T-3	2
103	Tube-Reinforcing	$5/8 \times .035 \times 4^{3}/_{8}$	AN-T-4	1
104	Tube-Sta. 29–T, 29–T'	$7/8 \times .049 \times 6^{3}/_{8}$	AN-T-3	2
105	Bracket-Torque Tube Bearing	$1^{1}/_{2} \times .065 \times 3^{25}/_{64}$	AN-S-11	2
106	Tube-Sta. 20–20'	$3/4 \times .049 \times 27^{3}/_{4}$	AN-T-3	1

INSPECTION

3. Check for:

    a. Tears, cuts, cracks, any signs of failure, or any damage in the fabric, wood, or metal covering
    b. Broken or damaged structural members that may be shown by distortion of the fabric, plywood, or metal covering
    c. Loose objects which might foul controls or cause damage by bouncing around
    d. Signs of leaks
    e. Condition of the protective coating and finish.

4. Check for proper upkeep, such as cleanliness, and proper attention to protective measures.

5. Particular attention should be paid to engine-mount fittings, wing-attachment fittings, landing-gear fittings, tail-surface-attachment fittings, and the proper attachment of any accessories within the fuselage or hull. These fittings should be examined for evidences of looseness, elongated boltholes, corrosion, or any evidence of failure, particularly in welded joints.

Wings. While the wings of most airplanes are similar in construction, they may be constructed of many different materials. The wing may be constructed entirely of metal with a metal skin, or of a combination of wood and metal, or largely of wood with a fabric or plywood covering. On many light aircraft, the spars and main supporting members may be of wood, the ribs of metal, and the wing covered with fabric. The wing may be of cantilever construction being entirely supported from within, but on most light aircraft, external supports in the form of struts or brace wires are used. In inspecting a wing, the procedure listed below should be followed.

Fig. 20. Fuselage Frame Diagram. (Key to Fig. 20).

REF. NO.	NAME	SIZE (IN INCHES)	MATERIAL	NO. PER ASSEMBLY
107	Brake and Rudder Pedal Mounting Right			1
108	Tube-Sta. 19–20′	$5/8 \times .035 \times 29\frac{1}{2}$	AN–T–4	1
109	Tube-Sta. 18–18′, 19–19′	$7/8 \times .049 \times 27\frac{3}{4}$	AN–T–3	2
110	Tube-Rear Seat Truss Outer Diagonal	$3/4 \times .049 \times 7$	AN–T–3	2
111	Tube-Sta. 18–19′	$5/8 \times .035 \times 33\frac{1}{8}$	AN–T–4	1
112	Tube-Rudder Guard Support	$5/8 \times .035 \times 6\frac{1}{32}$	AN–T–4	2
113	Tube-Sta. 30–30′	$7/8 \times .035 \times 16$	AN–T–3	
114	Slide Assy.-Seat Adjustment-Right			1
115	Tube-Sta. 29–31, 29′–31	$3/4 \times .049 \times 5\frac{7}{8}$	AN–T–3	2

AIRCRAFT MAINTENANCE AND SERVICE

Fig. 21. All metal construction of a light airplane wing. (Courtesy Engineering and Research Corporation)

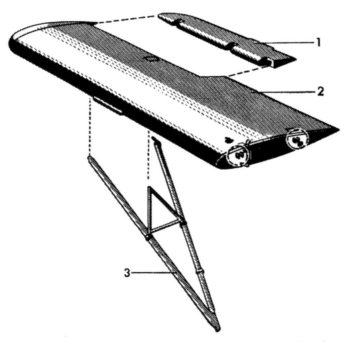

Fig. 22. A wing group showing (1) aileron assembly, (2) wing panel, (3) strut installation with lift and jury struts. (Courtesy Taylorcraft Aviation Corporation)

INSPECTION

1. The wings and center section should be checked for:
   a. Loose rivets or rivets pulling through the metal, the rib stitching pulling through the fabric, and the condition of other fastenings designed to hold the skin in place
   b. Broken or distorted ribs
   c. Drain grommets to make certain the opening is clear
   d. Tightness of internal brace-rods, wires, or wooden braces
   e. Condition of finish and aircraft markings
   f. Corrosion of metal parts and protective covering of wooden parts within the wing
   g. Condition of walkways for signs of loosening or wear.
2. Check hinges, pulleys, and other moving parts for proper lubrication.
3. Check slots for free and full movement.
4. Check wing terminals for:
   a. Security of attachment and elongated boltholes
   b. Cracks, checks, or signs of failure and proper safetying
   c. Any looseness in fittings.
5. Check flotation compartments, if present, removing drain plugs to allow any condensation to drain out.

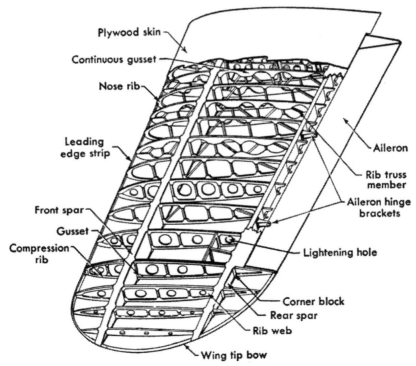

Fig. 23. All wood wing construction. (Courtesy Forest Products Laboratory)

AIRCRAFT MAINTENANCE AND SERVICE

6. Check general condition of fabric or metal covering and finish.

7. Check for any misalignment or distortion of any of the wing parts.

8. Check inspection plates, inspection apertures, and openings in skins through which push rods or cables pass.

Tail Group or Empennage. The tail group consists of the vertical fin, the horizontal stabilizer, and the rudder and elevators. (The rudder and elevators will be discussed in the next section under Flight-Control Surfaces.) Depending upon the type of aircraft, there may be one or more tail groups. The tail group may be constructed of wood, or of metal covered with fabric, or it may be entirely of metal. Most tail groups have some form of external bracing, although on major aircraft the structures may be self-supporting. In making the inspection of the tail group, the procedure given below should be followed.

1. Check the surfaces for:
    a. Loose or pulled rivets, stitching, or any other fastening designed to hold the skin in place
    b. Any signs of distortion or misalignment
    c. Condition of the finish or protective coating.

2. Check all braces, wires, and attachment fittings for:
    a. Any looseness, paying particular attention to proper safetying
    b. All fittings for signs of wear, cracks, or checks, particularly at welded joints
    c. Elongated boltholes and hinges for tightness
    d. Nicks or corrosion on the surface of tie rods, braces, wires, or fittings.

3. Check the general condition of:
    a. Fabric or metal covering, particularly the leading-edge parts
    b. Ribs for any signs of failure or displacement
    c. All bracing for security of attachment and proper tension.

Flight-Control Surfaces. The flight-control surfaces consist of the ailerons, elevators, rudder, and various trimming tabs. Flaps and spoilers also may be considered flight-control surfaces. The construction of control surfaces is usually as light as possible to include the desired strength. On many all-metal or all-wood airplanes, the control surfaces are of light wood or tube construction covered with fabric. The inspection of the flight-control surfaces should include the following items.

1. Check the horns and hinges for:
    a. Bends, breaks, or any signs of cracking or wear

INSPECTION

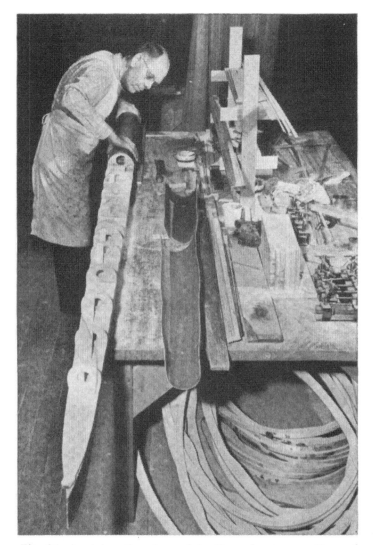

Fig. 24. Applying the plywood leading edge to an all wood elevator. (Courtesy Timm Aircraft Corporation)

    b. Security of all attachments
    c. Safetying and loose hinge pins.
2. Check braces and wires for:
    a. Proper tension
    b. General condition
    c. Safetying and security of attachments.
3. Check ailerons, rudder, elevators, tabs, and landing flaps for:
    a. Free and full movement

AIRCRAFT MAINTENANCE AND SERVICE

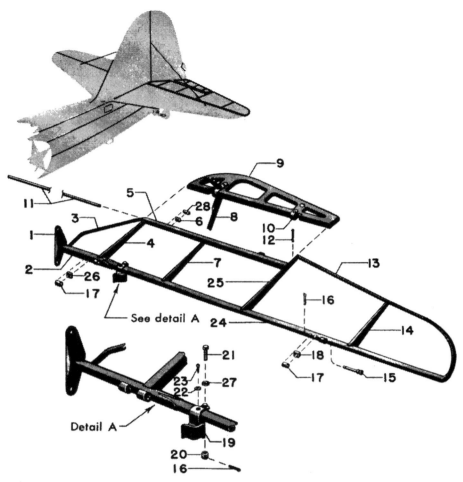

Fig. 25. Left elevator and trim tab metal frame assemblies showing (1) elevator control horns, (2) elevator-horn reinforcement sleeve, (3) butt trailing edge. (4) rib No. 1, (5) elevator cut-out tube, (6) filler block, (7) rib No. 2, (8) trim-tab hinge assembly, (9) elevator tab frame, (10) tab hinge assembly, (11) hinge pins, (12) cotter pin, (13) trailing edge, (14) rib No. 4, (15) tail-surface hinge pin, (16) cotter pin, (17) sleeve, (18) filler block, (19) trim-tab cable stabilizer clamp assembly, (20) shear nut. (21) aircraft bolt, (22) lockwasher, (23) screw, (24) elevator leading-edge tube, (25) rib No. 3, (26) filler block, (27) lock washer, (28) sleeve. Courtesy Taylorcraft Aviation Corporation.

   b. Warping or distortion of any kind
   c. Broken ribs or failure of any internal parts
   d. Condition of covering and finish
   e. Condition of grommet holes in fabric covering
   f. Proper operation of the tabs and their controls, and absolute security of their fastening.

INSPECTION

4. Check all surfaces for:
   a. Holes or other visible damage
   b. General condition
   c. Loose rivets, tears around rivets, or indication of failure of stitching in fabric.

5. Check all control surfaces for the security of the attachment of the hinges, brackets, control horns, torque tubes, and cables.

6. Remove or disengage all surface control locks.

7. Apply all surface control locks.

Fig. 26. Applying plywood skin to an all wood aileron. (Courtesy Timm Aircraft Corporation)

NOTE: Control surfaces should be inspected after they have been moved violently by wind or propeller blast during taxiing or after windy conditions while the airplane is parked.

Engine Mounts and Nacelles. Engine mounts not only support the engine, but also receive all forces transmitted to the aircraft from the engine. The mounts are subject to torsional and vibration strains. The mounts should be given careful inspection at frequent intervals. The inspection procedure given below should be followed.

AIRCRAFT MAINTENANCE AND SERVICE

Fig. 27. A typical engine installation showing cowling opened. (Courtesy Engineering and Research Corporation)

1. Inspect the entire engine-mount assembly for:
   a. Cracks or signs of failure, particularly in welds
   b. Tightness and safetying of mounting clamps and bolts
   c. Bent or broken structural members
   d. Proper condition of protective coating.

2. Inspect for proper condition of engine-mount vibration absorbers, such as shackles or pads.

Landing Gear. The landing gear is subjected to more abuse than perhaps any other part of the aircraft. Any poor landing subjects this structure to tremendous strains. The landing gear may be of the fixed or retractible type, but it always receives the full force of a bad landing. In inspecting the landing gear, the procedure given below should be followed.

INSPECTION

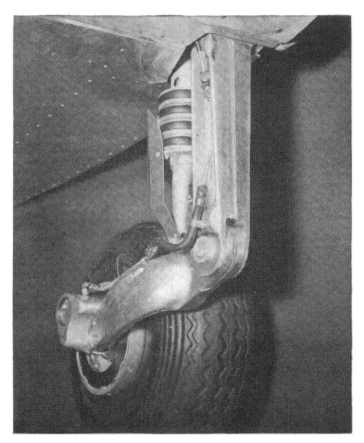

Fig. 28. A main landing gear wheel on a light airplane. Note rugged construction, shock absorbing disks, brake drums and flexible hydraulic connections. (Courtesy Engineering and Research Corporation)

1. Check the condition of:
   a. Ball and socket joints
   b. Shock absorber cord, if used.
2. Check oleo struts for proper fluid level.
3. Test for proper functioning of landing-gear retraction and lowering mechanism.
4. Check the condition and functioning of valves, rivets, pumps, pulleys, drums, torque tubes, pressure tubes, flexible connections, couplings, latches, stops, and limit switches. Also check the tension of the connecting and operating cables.
5. Check the operation of the landing-gear position indicators to see that they correctly indicate the position of the landing gear.

AIRCRAFT MAINTENANCE AND SERVICE

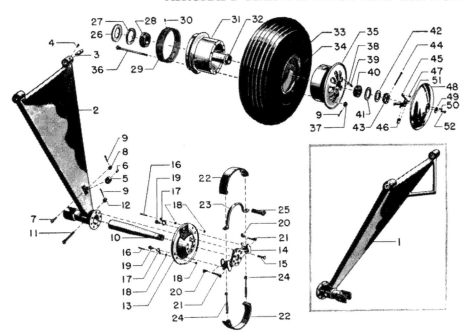

Fig. 29. A main landing-gear assembly showing (1) V-assembly, (2) V-assembly, (3) Landing-gear bushing, (4) stove head screw (5) cable bushing, (6) Oilite bushing, (7) aircraft bolt, (8) shear nut, (9) cotter pin, (10) landing-gear axle, (11) aircraft bolt, (12) castle nut, (13) brake dust-shields assembly, (14) anchor-plate assembly, (15) machine screw, (16) cotter pin, (17) clip, (18) washer, (19) shoe adjusting nut, (20) shoe adjusting wedge, (21) shoe adjusting wedge bolt, (22) brake-shoe assembly, (23) cam arm, (24) brake-shoe spring, (25) cam arm link, (26) washer, (27) washer, (28) wheel bearing, (29) brake lining, (30) brake-lining rivet, (31) inner half of wheel, (32) axle shield tube, (33) tire, (34) tube, (35) outer half of wheel, (36) assembly bolt, (37) castle nut, (38) valve-stem nut, (39) valve cap, (40) wheel bearing, (41) washer, (42) washer, (43) shear nut, (44) cotter pin, (45) tripod, (46) tripod nut, (47) tripod screw, (48) hub cap, (49) washer, (50) steel washer, (51) rivet washer, (52) screw. (Courtesy Taylorcraft Aviation Corporation)

6. Check the functioning and general condition of the warning lights, buzzers, and other warning devices.

7. Check the shock absorber units for permanent set or breakage of taxiing springs.

8. Check the rubber disks for:
   a. Condition of the rubber
   b. Distortion, stretching, or twisting, and to see that the rubber is free of oil.

9. Inspect the struts, braces, and fittings for:
   a. Cracks, particularly at bends or welded joints
   b. Any bends or distortions

INSPECTION

Fig. 30. A main landing gear wheel on a light airplane. Note wheel bolts and the construction of the center section where the wing is to be attached. (Courtesy Engineering and Research Corporation)

    c. Condition of attachments, their security and safetying
    d. Elongated boltholes or signs of wear
    e. Loose, missing, or unsafetyed bolts, nuts, and cotter pins.

10. Check all parts that require oiling for proper lubrication.
11. Check the brace wires for proper tension, security, and safetying.
12. Inspect for the general condition of:
    a. Struts, braces, and fittings
    b. Shock units as indicated by bottoming of struts
    c. Evidence of leakage of fluid or pressure in hydraulic or similar units.
13. Check to make sure that the landing gear is latched in the down position and remove the landing-gear safety lock pins, if used.

AIRCRAFT MAINTENANCE AND SERVICE

14. Be sure that the lock pins are removed prior to each flight.
15. After each flight, be sure that the safety lock pins are inserted.

Landing-Gear Wheels. The landing-gear wheels include the main landing-gear wheels and the nose or tail wheels. The wheels are an important part of the landing gear, and the failure of a wheel on take-off or landing may cause serious accidents. When inspecting the landing-gear wheels, the procedure given below should be followed.

1. Remove the nose or tail wheel and check:
   a. Bearing cups for scores, corrosions, or pits
   b. Entire wheel assembly for cracks, corrosion, condition of protective covering, dents in rims or disks, and loose wheel rivets, or any sign of failure
   c. Grease retainers for excessive wear, distortion, or grease soaking.
2. Remove the landing wheels and check:
   a. Brake drums for cracks, scores, and pits
   b. Bearing cups for scores, corrosion, and pits
   c. Grease retainers for excessive wear, distortion, or grease soaking
   d. Entire wheel assembly for cracks in wheel castings, corrosion, condition of protective coating, dents in rim disks, loose wheel rivets, bolts, nuts, studs, safetying, and any distortion or signs of failure.
3. Check wheels for:
   a. Adjustment and excessive wear between wheel bushing and axle
   b. Any excessive side-play or shake
   c. Snug fit of valve-stem washers on wheel.
4. At frequent intervals, all wheels should be checked visually for:
   a. Cracks, breaks, bends, and distorted flanges
   b. Condition of grease retainers and signs of grease leakage
   c. Condition of protective coating, security of retaining nuts, bolts, cotter pins, and lock rings
   d. Freedom from grease, mud, grass, and for proper lubrication.
5. Check the tires for:
   a. Tread wear
   b. External breaks, cuts, blisters, or any other visible damage
   c. Snugness of fit of valve stem when casing is mounted on wheel
   d. Any indication of rim damage to the casing
   e. Proper inflation.
6. Examine the wheel bearings for:
   a. Proper clearance between bushings and axles
   b. At frequent intervals, clean, lubricate, and inspect bearings for roughness and corrosion of races, pitted cups, out-of-round, pitted, or corroded rollers.

Fig. 31. Insulating the cabin of a four-place airplane with fiber glass blankets. (Courtesy Consolidated Vultee Aircraft Corporation, Stinson Division)

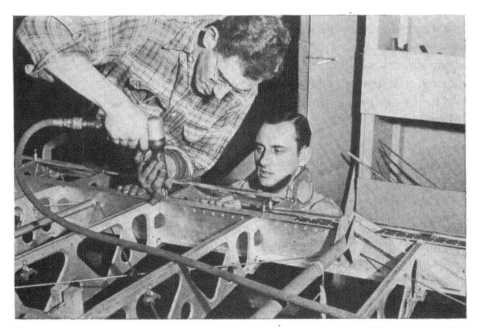

Fig. 32. Riveting the all-metal frame of a light, postwar airplane. Note construction of spars, ribs, compression braces and the installation of the drag and anti-drag wires. (Courtesy Consolidated Vultee Aircraft Corporation, Stinson Division)

AIRCRAFT MAINTENANCE AND SERVICE

7. The tail or nose gear should be checked for:
   a. Struts for proper fluid level
   b. Entire assembly for cracks, breaks, bends or signs of failure
   c. Weak or broken shock-absorber springs
   d. Condition of bearing post
   e. Condition of attachment fittings
   f. Operation of swivel relay mechanism
   g. Proper lubrication
   h. Condition of shock unit.

NOTE: At frequent intervals, the nose and tail-gear assemblies should be checked for freedom from mud, grass, and accumulation of foreign matter, and for worn or loose shoes if a tail skid is used. These assemblies should be given a visual inspection previous to each flight.

Wheel Brakes. There are a number of different brake systems in common use. The inspector or mechanic should obtain the manufacturer's service bulletins for the type of brake concerned and follow the recommended procedure for maintenance and service.

Cowling and Fairing. Cowlings and fairings are not usually classified as structural members. Cowlings enclose the engine, decreasing air resistance by their streamlined shape. Cowlings are designed to assist in engine cooling.

Fairings have the function of streamlining. Fairings are used to blend various structural parts of the airplane into other structural parts to allow smooth air flow. Fairings may be of light wood, such as balsa, or properly formed light sheets of metal.

The inspection of the cowlings and fairings given below should be followed.

1. Check for cracks, defective fasteners, condition of padding, and protective coating.
2. Check cowling, support-cups, and brackets for the condition and security of the attachments.

(The following should be performed daily or before each flight)

3. Check daily or before each flight for general condition, dents, bends, and security of attachments.
4. Check daily or before each flight for security of all cowling doors and covers.

(The following should be performed after each flight)

5. Remove enough cowling after each flight to check for:
   a. Fuel and oil leaks within the engine nacelle

INSPECTION

  b. Failure of wires, lines, connections, or fastenings
  c. Attachment of all exhaust pipes, collector rings, baffles, air scoops, and other accessories.

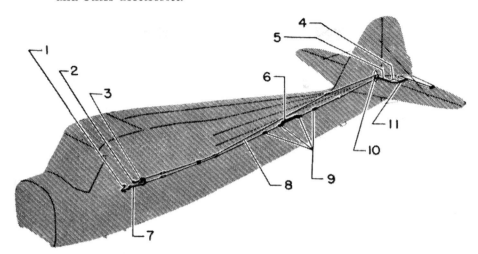

Fig. 33. The elevator trim tab control system showing (1) trim tab indicator wire, (2) trim tab control handle, (3) trim tab control pulley (front), (4) trim tab bell-crank assembly, (5) trim tab drive assembly, (6) trim tab cable spring, (7) trim tab indicator-arm assembly lever, (8) trim tab control cable, (9) trim tab cable clamp, (10) trim tab control rear pulley, (11) trim tab link assembly. (Courtesy Taylorcraft Aviation Corporation)

Flight-Control Mechanism. The flight-control mechanism consists of the stick or wheel-control assembly, the rudder bars, pedal assemblies, tab and landing-flap controls. The control mechanism includes all cables, pulleys, torque tubes, push or pull rods, and any other connection between the cabin controls and the flight-control surfaces. Inspection procedure of the flight-control mechanism is given below.

1. Check for the proper tension of the control cables.
2. Clean and treat all cables with rust-preventive compound.
3. Check the rudder bar and pedal assembly, stick or wheel-control assembly, tab and landing-flap mechanism for:
   a. Condition and functioning of parts
   b. Cleanliness and condition of post bearings and pedal bearings
   c. Position of control surfaces relative to the position of the control stick or wheel and rudder pedals
   d. Check for the correct position of flaps and flettners on cockpit indicators
   e. Security of control sticks in sockets
   f. Free and full movement of all control surfaces.

AIRCRAFT MAINTENANCE AND SERVICE

4. Check for:
   a. Condition of cables, chains, rods, turnbuckles, pulleys, brackets, fairleads, guides, attachment fittings, and bonding
   b. Lost motion
   c. Security of attachment and proper safetying.

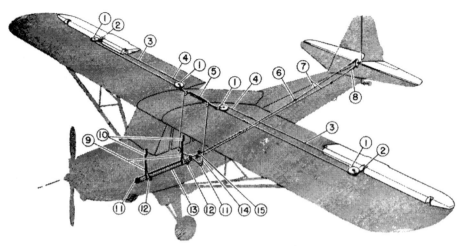

Fig. 34. The aileron and elevator control system showing (1) aileron cable pulley, (2) aileron control fittings, (3) aileron cables, (4) aileron cable attachments, (5) bell crank assembly, (6) elevator cable cramps, (7) elevator controlled cables, (8) turnbuckles, (9) stick attachments, (10) sticks, (11) pulleys, (12) torque pulleys, (13) torque tube, (14) cables, (15) pulleys. (Courtesy Taylorcraft Aviation Corporation)

5. Check landing-flap, hydraulic operating mechanism for:
   a. Leaks
   b. Kinked or damaged pipes or connections
   c. Proper operation.

6. Check to see that all parts are properly lubricated as required.

7. Check for freedom of operation of all controls previous to flight.

Wing Struts and Bracings. Wing struts and bracings should be checked for:

a. Bent or bowed struts or brace tubes
b. Proper tension of cables and streamline wires
c. Cracks, checks, dents, or any signs of failure
d. Condition and security of all bolts, nuts, fittings, turnbuckles, and safetying devices.

NOTE: Wing struts and bracings should be given visual checks for general condition, security of attachments, proper alignment, and tension previous to flight.

INSPECTION

Bonding. Bonding is a means of preventing radio interference by making the proper electrical connections between all parts of the aircraft. The entire bonding system should be checked at frequent intervals to determine the condition of the bonding braid and connections. The bonding braid should be inspected at the connections for signs of corrosion.

DAILY FLIGHT INSPECTION RECORD

DATE_____

Plane_____Operator_____
    (Make and Model)

This form should be completed daily as indicated for each plane operated, and placed on a clip board to remain with the aircraft during the day's operations.

PART I. LINE INSPECTION

Complete before starting flight. Check satisfactory items; no others.

Check Column

A. PROPELLER.
    1. Inspect BLADES for pits, cracks, nicks _____ _____

    2. Inspect HUBS and ATTACHING PARTS for defects, tightness,

       and safetying _____ _____

    3. Check PROPELLER for track _____ _____

B. ENGINE.
    1. Inspect ENGINE COWLING for cracks and security _____ _____

    2. Inspect EXHAUST STACKS and COLLECTOR RING for cracks

       and security _____ _____

    3. Check VALVE-GEAR MECHANISM and LUBRICATE as necessary _____

    4. Check SPARK-PLUG TERMINAL ASSEMBLIES for cleanliness

       and tightness _____ _____

5. Inspect accessible IGNITION WIRING and HARNESS for security of mounting _____ _____

6. Clean main FUEL-LINE STRAINERS _____ _____

7. Drain small quantity of fuel from bottom drain and inspect _____ _____

8. Check FUEL and OIL SYSTEMS for leaks, vent openings, and fit of tank caps _____ _____

9. Check FUEL and OIL supply (do not rely on gauges) ____ _____

10. Check all BOLTS and NUTS on engine and mount _____ _____

11. Turn propeller; check COMPRESSION of CYLINDERS _____ _____

C. LANDING GEAR.
1. Inspect TIRES for defects and proper inflation _____ _____

2. Inspect SHOCK-ABSORBER UNITS and BRAKE-LINKAGE GEAR _____

3. Inspect WHEELS for cracks and distortion and HUB CAPS for security _____ _____

4. Inspect STRUT-RETAINING BOLTS and FITTINGS for security _____

5. Inspect BRACE WIRES for tension and security _____ _____

6. Inspect MAIN FLOAT(s) for leaks and security of handhole covers _____ _____

D. WINGS.
1. Inspect COVERING for damage, buckled ribs, and end bows _____

2. Inspect ATTACHMENT FITTINGS for security _____ _____

3. Check STRUTS and FLYING WIRES for security of terminal connections _____ _____

4. Check AILERON HINGES, PINS, HORNS, and TABS _____ _____

INSPECTION

    5. Inspect accessible CONTROL CABLE, TUBES, and PULLEYS for security _____ _____

E. TAIL CONTROL SURFACES.
    1. Inspect COVERING for damage, buckled ribs, and bruised edges _____ _____

    2. Inspect ATTACHMENT FITTINGS for security _____ _____

    3. Check STRUTS and BRACE WIRES for security of terminal connections _____ _____

    4. Check CONTROL-SURFACE HINGES, PINS, HORNS, and TABS _____

    5. Inspect CONTROL CABLE, TUBES, and PULLEYS for security and lubrication _____ _____

    6. Check STABILIZER ADJUSTMENT assembly mechanism _____ _____

    7. Check TAIL-SKID or WHEEL assembly for condition and lubrication _____ _____

F. FUSELAGE.
    1. Inspect COVERING for damage and distortion _____ _____

    2. Inspect CONTROL-COLUMN assembly and accessible parts of control system for freedom of movement and security of attachments _____ _____

    3. Inspect RUDDER-PEDAL assembly and CONTROL SYSTEM as above _____ _____

    4. Check STABILIZER ADJUSTMENT mechanism for freedom of movement _____ _____

    5. Locate FIRE EXTINGUISHER and FIRST-AID KIT _____ _____

    6. Inspect all removable COWLING, FAIRING, and INSPECTION PLATES for security _____ _____

AIRCRAFT MAINTENANCE AND SERVICE

    7. Check proper functioning of LIGHTING SYSTEM _____ _____

    8. Inspect for security of SAFETY BELTS _____ _____

    9. Check functioning of ENCLOSURES and ADJUSTABLE-SEAT

        mechanism _____ _____

G. WARMING UP.
    1. See that CHOCKS are under wheels _____ _____

    2. Warm up and check proper functioning of ENGINE ____ _____

    3. Test engine (s) on each MAGNETO and on all TANKS ___ _____

    4. Check ENGINE CONTROLS for proper functioning and lost

        motion _____ _____

    5. Check position of CARBURETOR AIR PREHEATER _____ _____

    6. Check operation of CARBURETOR MIXTURE CONTROL ___ _____

    7. Check RADIO EQUIPMENT for proper functioning _____ _____

    8. Oil temperature _____ Oil pressure _____ R.p.m. _____

        Am't fuel _____ Am't oil _____ _____ _____

I CERTIFY that above airplane has this day been inspected under my supervision and that the aircraft is (is not) ready for flight.

Signed: _____ _____
                                                                        Supervisor of Inspection

PART II. SERVICING RECORD

Complete as necessary during and at finish of light operation.

	1st	2nd	3rd	4th	5th	
Gallons of gas						
Quarts of oil						

INSPECTION

PART III. REPORT AFTER FLIGHT

To be completed by each student after each flight.

Flight No.	Air time	Average r.p.m.	Oil pressure	Oil temperature	During flight I have noticed the following defects in this plane which should be remedied:	Student's initials
1						
2						
3						
4						
5						
6						
7						
8						
9						
10						
11						
12						
13						
14						
15						
16						
17						
18						
19						
20						

Total time _____ _____   Total gallons of gas _____ _____   Total quarts of oil _____ _____
Previous time this month _____ _____   Previous gallons this month _____ _____   Previous quarts this month _____ _____
Total this month _____ _____   Total this month _____ _____   Total this month _____ _____
Gallons of gas per hour consumed _____   Quarts of oil per hour consumed _____ _____

V GENERAL MAINTENANCE OF THE AIRPLANE

It is of the greatest importance in aircraft maintenance that the smallest failure be found and corrected before it becomes serious. Small cracks are indications of failure of either the wooden or metal structure and should be taken care of at once. The various parts of the airplane should be carefully examined to detect any indication of failure for, when failure occurs, there very often "isn't an airplane to be examined." Any complete failure of an airplane structural part may lead to the complete destruction of the airplane.

Engine Mount. The engine mount should be carefully examined for cracked, bent, dented, or broken members of the structure, even though there is no general indication of failure. The complete failure of any part of the engine mount creates a highly dangerous condition. Without exception, any part which shows any indication of failure should be immediately repaired or replaced before the airplane is flown.

Cracks are most likely to occur at welded points. These cracks may be difficult to discover because of protective coatings, such as paint, or accumulations of oil and dirt. The engine mount should be thoroughly cleaned at regular intervals, and special care must be taken in making inspections of this part of the aircraft. All mounting bolts and clamps should be tested for tightness and safety. The engine should be moved, if possible, on the mount by rocking up and down by one person while another person watches for any movement that might indicate elongated boltholes or looseness of any fastenings. If the protective coating is off, it should be immediately replaced to prevent rusting. If rust spots are discovered, the rust should be removed down to the clean metal and the protective coating renewed.

Fuselage and Wings. The fuselage and wings should be carefully checked for any evidence of corrosion either inside or out. Any distor-

GENERAL MAINTENANCE OF THE AIRPLANE

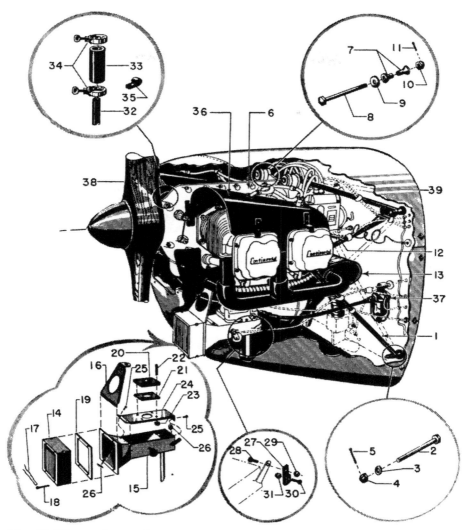

Fig. 35. The engine section group showing (1) Engine mount assembly, (2) aircraft bolt, (3) plain washer, (4) castle nut, (5) cotter pin, (6) engine dual ignition, (7) shock mount, (8) aircraft bolt, (9) washer, (10) castle nut, (11) cotter pin, (12) bonding assembly, engine to fuselage, (13) exhaust-system installation, (14) filter, (15) housing, (16) support, (17) lock wire, (18) screw, (19) gasket, (20) adapter, (21) gasket, (22) stud, (23) nut, (24) cover and flange, (25) screw, (26) speed nut, (27) carburetor heater control clamp, (28) roundhead machine screw, (29) self-locking nut, (30) round-head machine screw, (31) self-locking nut, (32) oil breather tube, (33) oil breather-tube hose, (34) hose clamp, (35) breather-tube clamp, (36) engine baffle installation, (37) fuel-system installation in engine compartment, (38) propeller installation, and (39) cowl assembly. (Courtesy Taylorcraft Aviation Corporation)

AIRCRAFT MAINTENANCE AND SERVICE

tion or wrinkling of the surface covering of the fuselage or wings may indicate internal distortion or damage to internal parts. Corrosion is most likely to occur in pockets or corners where moisture may collect. Grommets and drain holes should be kept open and clean. Damage

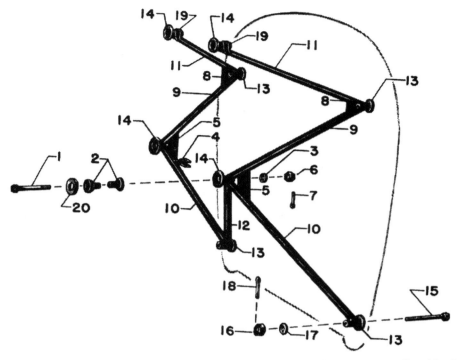

Fig. 36. An engine mount showing (1) aircraft bolt, (2) shock mounts (furnished with engine), (3) washer, engine to mount (furnished with engine), (4) throttle-control bracket assembly, (5) engine-mount gusset, (6) castle nut, (7) cotter pin, (8) gusset, (9) (10) (11) and (12) tubes, (13) and (14) sleeve assembly, (15) aircraft bolt, (16) castle nut, (17) plain washer, (18) cotter pin, (19) strap, and (20) washer (furnished with engine). (Courtesy Taylorcraft Aviation Corporation)

caused by contact with stones which have torn a hole in the covering is easily discovered, but a broken wing or distorted part of the internal structure may be indicated only by wrinkles in the fabric or metal skin, by a buckling of the metal covering, loose rivets, or by various parts being out of line. The outside of the structure should be carefully examined for any indications of internal injury, distortion, or failure.

Warped wings are usually indicated by parallel wrinkles running diagonally across the wing over a comparatively large area. These wrinkles may appear in either the fabric or the metal skin. Warped wings may be caused by unusually violent maneuvers, extremely rough

GENERAL MAINTENANCE OF THE AIRPLANE

air, or hard landings. There may be no actual destruction of any part of the structure, but it may be distorted and weakened to such an extent that failure may result later. Such indications may also occur on the surface of the fuselage.

Any wrinkling of the covering should lead to a careful examination of the internal parts. Small cracks or tears in the wing covering should

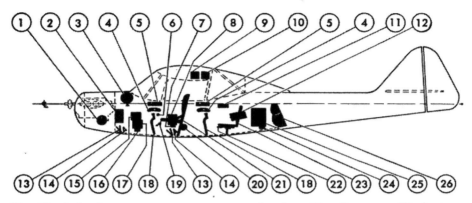

Fig. 37. A fuselage contents arrangements showing (1) oil sump, (2) battery, (3) collector tank, (4) throttles, (5) trim-tab controls, (6) throttle friction control, (7) radio control box, (8) engine control panel, (9) radio receiver, (10) radio transmitter, (11) first aid kit, (12) mapholder, (13) rudder pedals, (14) brake pedals, (15) radio receiver, (16) junction box, (17) radio transmitter, (18) control sticks, (19) front seat adjustment handle, (20) fire extinguisher, (21) front seat, (22) rear seat adjustment handle, (23) rear seat, (24) dynamotor, (25) baggage compartment, and (26) baggage hold-down strap. (Courtesy Taylorcraft Aviation Corporation)

be checked and repaired at once. Small cracks, such as those leading away from rivets, frequently occur in metal-covered airplane wings or fuselage. A small hole about $1/32$ in. in diameter may be drilled at the end of the crack to stop the crack from continuing to develop. A permanent repair should, however, always be made by riveting on a patch.

Tearing out along the edge of a rib or a seam on a fabric-covered structure is usually an indication of some giving away of internal parts. A failure of this kind should be repaired in an approved manner after it is determined that there is no internal failure or after the internal failure has been repaired.

Aluminum sheets are usually made of an alloy of aluminum which is heat-treated, and they are springy and hard to bend. No attempt should be made to remove dents or straighten bent parts. The injured material should be removed and replaced by a properly riveted patch.

AIRCRAFT MAINTENANCE AND SERVICE

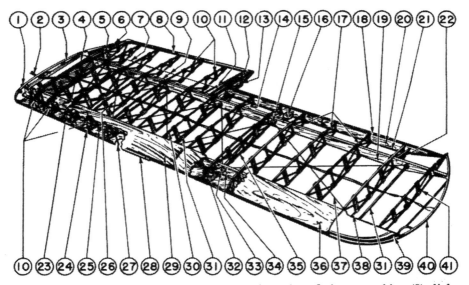

Fig. 38. A wing-frame assembly showing (1) front butt-fitting assembly, (2) light wire assembly, (3) No. 1 rib assembly, (4) fuel-tank installation, (5) aileron control-pulley and rear butt-fitting installation, (6) fairing assembly, butt, (7) inner aileron cable assembly, (8) trailing-edge assembly, (9) outer aileron cable assembly, (10) rib assembly, (11) diagonal rib brace, (12) upper rib brace assembly, (13) lower rib brace assembly, (14) No. 9 rib assembly, (15) rear spar strut fitting assembly, (16) aileron control bracket installation, (17) rib assembly, (18) trailing-edge assembly, upper aileron cutout, (19) rear spar assembly, (20) aileron cutout fairing installation, (21) trailing-edge assembly, lower aileron cutout, (22) aileron hinge bracket installation, (23) No. 4 rib assembly, (24) inboard leading-edge cover assembly, (25) jury strut spar attachment, (26) drag wire, (27) pitot static tube assembly, (28) spin strip, (29) center leading-edge cover assembly, (30) No. 8 rib assembly, (31) drag strut installation, (32) No. 10 rib assembly, (33) front spar strut fitting assembly, (34) front spar assembly, (35) strut truss assembly, (36) outboard leading-edge cover assembly, (37) drag wire, (38) No. 14 rib assembly, (39) tip leading-edge cover assembly, (40) wing bow assembly, and (41) tip rib assembly. (Courtesy Taylorcraft Aviation Corporation)

Aluminum-coated alloys should be carefully examined to determine whether or not cuts, scratches, or small injuries have penetrated the pure aluminum covering. The alloy underneath is apt to corrode rapidly unless protected from the elements. Small spots or injuries should be covered with lacquer or enamel. If a large area is affected, the entire area should be replaced or refinished with a spray gun in an approved manner.

Any damage to the protective covering of an aircraft should be repaired at once.

Control Surfaces. Movable parts, such as control surfaces and flaps, should be checked to see that the hinges are in good condition and all

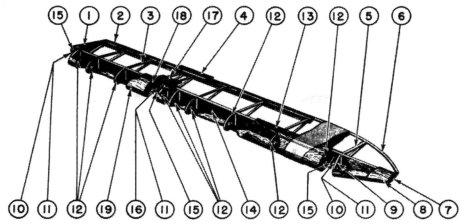

Fig. 39. An aileron-frame assembly showing (1) wood aileron butt, (2) wood aileron trailing-edge assembly, (3) main wood aileron rib assembly, (4) aileron tab, (5) end wood aileron rib assembly, (6) trailing-edge wood aileron bow assembly, (7) tip wood aileron rib, (8) tip wood aileron leading edge, (9) tip bulkhead wood aileron rib assembly, (10) aircraft bolt, (11) self-locking nut, (12) bulkhead wood aileron rib assembly, (13) center wood aileron leading edge, (14) wood aileron spar assembly, (15) aileron bracket assembly, (16) aircraft bolt, (17) control point gusset, (18) bulkhead wood aileron rib assembly, and (19) inner wood aileron leading edge. (Courtesy Taylorcraft Aviation Corporation)

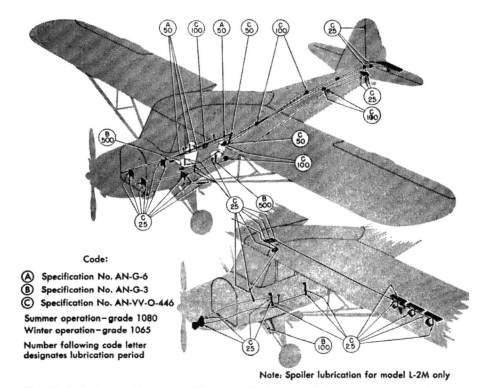

Fig. 40. Lubrication Diagram. (Courtesy Taylorcraft Aviation Corporation)

AIRCRAFT MAINTENANCE AND SERVICE

movable bearing parts are properly lubricated. When cleaning with solvents, metal-shielded ball or roller bearings should be protected to prevent the grease in the bearing from being washed out. These parts are usually equipped with bearings which are packed with grease at the time of assembly and usually do not need to be relubricated except at major overhaul periods.

Control Cables. The control cables leading to movable parts should be adjusted for proper length so that control elements, such as the rudder pedals and the stick, will be in the neutral position when the control surface is in its neutral position. The turnbuckles which are used to adjust the length of the control cables should always be examined for proper safetying. Control cables are usually made up of a number of wires and do not usually fail all at once. The cables should be carefully examined to determine any fraying which indicates broken individual wires. This condition should be looked for where the cables make an abrupt turn, such as over a control-cable pulley. Any frayed condition is reason for replacement of the cable.

Control Pulleys. Control pulleys should be examined for binding, worn or damaged grooves, and to see that they rotate freely on their bearings. The fastening of the pulley should be examined for looseness and proper safetying. Any indication of rust on a control cable should be carefully examined for indications of internal damage which, if found, is an indication that the cable should be replaced. Care should be taken to see that there is no object in contact with the control cables which might cause jamming. The control cable should also be protected from any rubbing against articles stored in the airplane.

Landing Gear. The main landing gear and the tail gear should be kept in perfect condition. At regular intervals the landing gear should be examined for any signs of failure. Landing-gear bracings and shock-absorbing equipment should be checked daily, or before flight.

At regular intervals, airplanes equipped with retractable landing gear should be placed on jacks and all parts operated through complete cycles by means of the controls. As the landing gear is extended and retracted, a careful check should be made to see that it operates freely throughout its total range. Be sure that no part binds on any part of the airplane structure.

When shock cord is used on a landing gear, it should be inspected for broken and fraying covering or any signs of breaks in the rubber cords.

GENERAL MAINTENANCE OF THE AIRPLANE

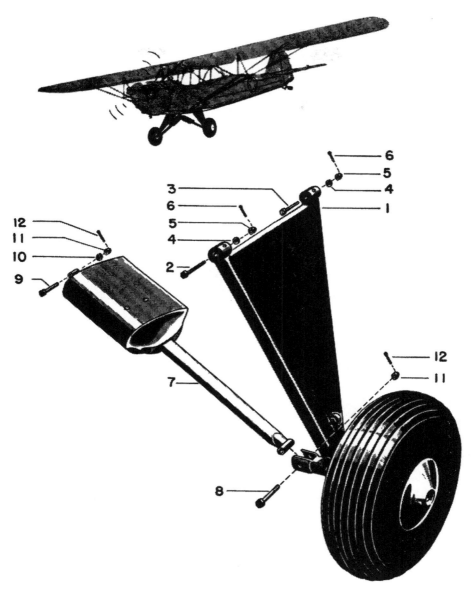

Fig. 41. A landing-gear installation showing (1) landing-gear assembly, (2) landing-gear tie strut assembly, (3) castle nut, (4) castle nut, (5 and 6) cotter pins, (7, 8, 9 and 12) aircraft bolts, (10) cadmium plated washer, (11) washer. (Courtesy Taylorcraft Aviation Corporation)

AIRCRAFT MAINTENANCE AND SERVICE

The shock cords should be kept free from grease, because grease causes rapid rotting of rubber.

Landing-gear cables and other cables should be examined for proper tension and signs of fraying. Frayed cables should be replaced. The locking device on retractable landing gear should operate perfectly.

If the airplane is equipped with a landing-gear position indicator, it should be checked to see that it shows the true position of the land-

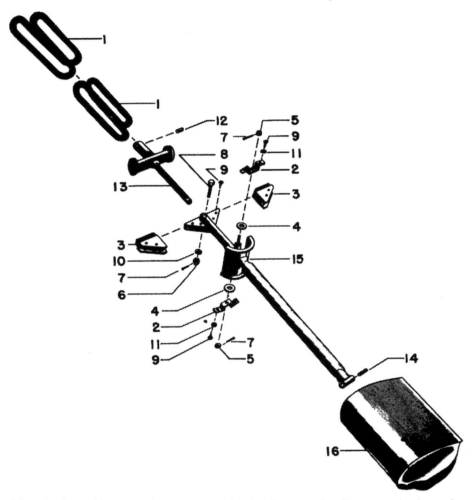

Fig. 42. A landing-gear tie-strut assembly showing (1) landing-gear shock cord, (2) shock cord boot support lug, (3) landing-gear tie-strut bumper, (4) shock cord retaining disk, (5) shear nut, (16) shear nut, (7) cotter pin, (8) aircraft (shock strut telescope) bolt, (9) Type A screw, (10) plain (telescoping bolt) washer, (11) plain (rivet burr) washer, (12) tie-strut upper bushing, (13) landing-gear upper-tie-strut assembly, (14) bushing, (15) landing-gear lower-tie-strut assembly, (16) shock cord assembly boot. (Courtesy Taylorcraft Aviation Corporation)

GENERAL MAINTENANCE OF THE AIRPLANE

ing gear, especially the retracted and extended positions. If it is equipped with a warning signal, the warning signal should be checked for proper operation and the signal should be clear and distinct whenever the landing gear is in any position other than down and locked.

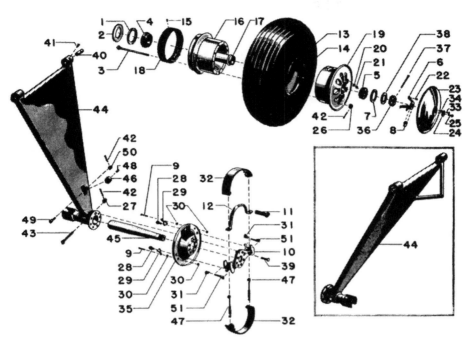

Fig. 43. A landing-gear assembly showing (1) bearing grease retainer, (2) bearing washer, (3) assembly bolt, (4) and (5) wheel bearing, (6) tripod, (7) bearing grease retainer, (8) tripod nut, (9) cotter pin, (10) anchor plate assembly, (11) long cable clip, (12) cam arm forging, (13) tire, (14) inner tube, (15) brake-lining rivet, (16) inner half of wheel assembly, (17) axle shield tube, (18) brake lining, (19) outer half of wheel assembly, (20) valve stem nut, (21) valve cap, (22) tripod screws, (23) hub-cap assembly, (24) washer, (25) roundhead machine screw, (26) and (27) castle nuts, (28) shoe adjustment nut, (29) adjustment nut lock spring, (30) spring lock washer, (31) shoe adjustment outer wedge, (32) brake shoe assembly, (33) rivet, (34) plain washer, (35) brake dust shield assembly, (36) shear nut, (37) cotter pin, (38) washer, (39) flat-head machine screw, (40) landing-gear bearing, (41) Type A screw, (42) cotter pin, (43) aircraft bolt, (44) welded (covered) V-assembly, (45) landing-gear axle, (46) cable (pulley, small steel) bushing, (47) release spring, (48) Oilite bushing, (49) aircraft bolt, (50) shear nut, (51) adjusting wedge bolt. (Courtesy Taylorcraft Aviation Corporation)

The tail landing-gear antishimmy device should be checked to see that it is free of grease and oil between the friction disks. The disks may be separated and washed with unleaded gasoline or other suitable solvent to remove oil and grease.

If the airplane is equipped with air-oil shock struts, they should be checked at regular intervals to see that the fluid is at the proper level, and fluid should be added if necessary. All air pressure in the strut should be relieved before checking the fluid level. Care should be taken in relieving the air pressure. Some plugs are fitted with a straight thread, and others with a tapered pipe thread. When the straight thread is used, the plug may be turned out gradually until the pressure-relief hole is uncovered, allowing the air to escape. If a tapered pipe thread is used, there is no vent in the side, and the plug, if unscrewed, will blow out and may cause injury. When the tapered plug is used, the air should be released from the strut by depressing the valve core. If the core is damaged, it should be replaced with a new one. The cores are similar to those used in automobile tires, but are especially designed with an oilproof seat. Ordinary tire valves should never be used.

The fluid level should be even with the filler-plug opening. Whenever the fluid level is below this point, additional fluid should be added. When filling a strut which is entirely empty of fluid or if the fluid is very low, care should be taken to make sure that all air is removed from the strut. After the strut has been filled with fluid to the plug opening, the plug may be inserted loosely and the strut extended and collapsed several times by raising and lowering the jacks. This operation should be continued until there is no change in the fluid level. Be sure the copper gasket is in place when the filler plug is screwed in tightly.

After filling with fluid and replacing the plug, the strut is filled with air from a high-pressure air source or with an air pump. The strut should be inflated in accordance with the manufacturer's instructions. If the strut is overinflated, a small quantity of air may be removed by carefully depressing the valve core. Care should be used that the air escapes slowly to avoid damage to the valve-core seat.

A small amount of oil should escape around the packing gland to lubricate the piston. Any appreciable amount of leaking should be stopped by tightening the packing gland. Air should not escape around the piston between it and the packing gland. The escape of air may be tested for by the use of soapy water. If it is necessary to tighten the gland, all pressure should be removed from the strut. If normal tightening does not stop the leak, the gland must be removed and the packing replaced. This is usually done in an approved shop or repair station. Alcohol is the only fluid suitable for cleaning or flushing out struts

GENERAL MAINTENANCE OF THE AIRPLANE

which have contained hydraulic fluid. Mineral oil will destroy the packing matter. Carbon tetrachloride forms an acid in contact with the hydraulic fluid.

Tires. The landing-wheel tires must be carefully inspected for **damage**. If the tread is cut deeply enough to expose the fabric, if there

Fig. 44. Landing gear. (Courtesy of Engineering and Research Corporation)

is any sign of breaks in the carcass of the tire, if large side-wall blisters appear, or if there is damage to the beads extending through the rubber covering to the fabric, the tire should be replaced. Particular attention must be paid to any indication of internal damage to the carcass of the tire. Each time a tire is removed, the inside of the tire should be carefully examined by the use of a tire spreader.

The wheel and rim should be thoroughly cleaned each time the tire is removed. If the protective covering is worn to such an extent that the metal of the wheel or rim shows through the covering, it should be renewed and the tire mounted in accordance with the manufacturer's directions.

Whether or not damage is repairable depends largely upon the good judgment of the mechanic or inspector. The inner tube should be replaced if there is any evidence of thin spots, chafing, areas which are damaged due to breaks in the carcass of the tire, or if there is any

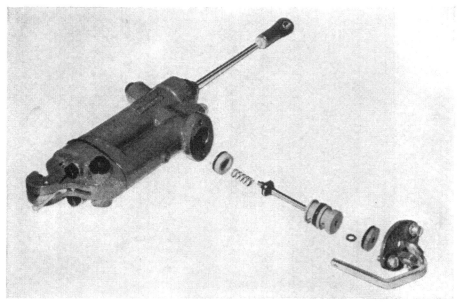

Fig. 45. Exploded view of the parking valve of a master brake cylinder. (Courtesy Gladden Products, Division Los Angeles Turf Club)

damage to the valves, or signs of failure of the valve attachment to the tube. The proper inflation of the tires should be checked daily or before flight.

Wheels. Every time a wheel is removed for inspection the entire surface of the wheel should be cleaned thoroughly. Bearings and gear retainers are removed and all grease removed with clean, unleaded gasoline. Compressed air may be used to dry the gear walls and blow out any foreign matter. The bearings should not be allowed to spin in the air blast. After cleaning, all surfaces should be inspected carefully for nicks, cracks, and battered places. All parts of the wheel, including hubs, webs, rims, flanges, and brake drums, should be carefully examined.

GENERAL MAINTENANCE OF THE AIRPLANE

A new wheel should be installed if loose liners, cracks, badly corroded areas, bent rims, or any distortion of the wheel itself is found. Particular care should be taken that the part of the wheel in contact with the tire is smooth and covered with the proper protective coating. After the bearings are thoroughly cleaned, they should be examined to see that the rollers are not pitted or corroded and the cover and cups are not out of round, cracked, rough, or showing any signs of failure.

If an arbor press is not available for removing the cups, they may be removed with a hammer and a soft metal bar, such as a bar of brass. Light blows should be used, and the bar should be moved around the

Fig. 46. Exploded view of master brake cylinder. (Courtesy Gladden Products, Division Los Angeles Turf Club)

cup one third of its circumference between blows. The proper lubricant should always be used on wheel bearings. Careful adjustment of the roller bearings must be made when installing the wheel. When the axle nut is tightened enough that the wheel bearing begins to drag when spinning the wheel by hand, the nut should be backed off to the next castellation and locked in place with the proper cotter pin. Care should be taken that brake drag is not confused with bearing tightness. Be sure that the brake is backed off far enough that there is plenty of clearance all around the drum between the drum and the brake shoes. Any replacements must be made with approved parts.

Brakes. During the inspection of the main landing-gear wheels, a careful check of the brakes and the brake assembly should be made.

The brake assembly should be renewed if any cracked, corroded, or broken parts are found. Distorted brake shoes, a badly scored brake lining, brake linings that are excessively worn or oil soaked, or loose rivets are cause for rejection of the brake assembly. Distorted brake shoes do not allow an even contact with the brake drum. "Grabby" brakes are commonly caused by grease-soaked shoes or brake lining. Unless the greasy condition is of comparatively long standing, the grease may be removed from the shoe or brake lining by the use of unleaded gasoline. After cleaning, the lining should be wiped clean with a dry cloth or a fine brush.

The return springs should be carefully inspected for rusting or signs of failure of the hook at each end of the spring. The spring should be replaced when there is any sign of failure.

The brake mechanism should be thoroughly cleaned. The cavity of the brake drum in the wheel and the bearing retainer should be thoroughly cleaned of grease or other foreign matter. Before adjusting the brakes, the wheel should be jacked up and the brake applied several

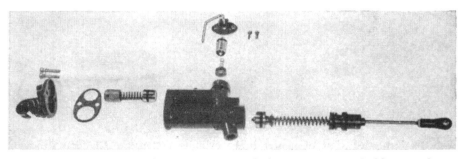

Fig. 47. Exploded view of master brake cylinder. (Courtesy Gladden Products, Division Los Angeles Turf Club)

times to be sure that the brake is operating properly and that it releases promptly. Cables or controls should be carefully examined, and any frayed or rusted cables should be replaced. Bell cranks and pulleys should be properly lubricated.

The adjustment of the brakes should be made in accordance with the manufacturer's directions and recommendations.

Cowling and Fairings. The cowling and fairings should be inspected regularly to see that they are not damaged and that their fastenings are secure. Care must be exercised in handling cowlings and fairings to prevent bending or distortion. Many pieces of cowlings and fairings are in the form of long strips of sheet metal which are not strong

GENERAL MAINTENANCE OF THE AIRPLANE

enough to support their shape unaided. Most of them are cut from light-gauge material which may be damaged easily.

Fairings and cowlings are subject to cracking. All cracks should be drilled at the end to prevent the extending of the crack and should be properly patched, or the cowling or fairing replaced. All fastenings must be kept in proper working condition and should be replaced at the first sign of looseness or excessive wear. The chafing strips between the various parts of the cowling which may rub together, should be renewed as soon as worn. Metal parts should never be allowed to rub together.

Accessories. All accessory parts of the main aircraft structure, such as seats, canopies, doors, windshields, windows, and safety belts, should be inspected at regular intervals and kept in good repair. Cracks in seats or sharp projections which might catch or tear the clothing or parachute should be repaired.

Transparencies. Windshields, windows, canopies, and other transparencies should be carefully sealed around the edges and kept clean. These parts may be cleaned with a very fine metal polish. Care should be taken not to rub hard enough to rub particles of sand or grit into the surface. The parts should be rinsed with clean water and wiped lightly with a soft cloth. Kerosene or naphtha may be used to remove oil or grease. The padding that holds the glass or plastic in place should be kept in good condition to prevent cracking due to vibration. Safety belts and their fastenings should also be kept in good condition. Door hinges and fastenings should be examined carefully, and any quick-release mechanism should be checked at regular intervals to see that it is operating properly.

Cleaning Compounds. It is the responsibility of the ground crew to keep the airplane clean at all times, both inside and out. Cleaners are divided into two general groups: those made up of water containing cleaning materials, and those made up of solvents which are volatile and often inflammable. Whenever possible, the water-solvent cleaner should be used. This type of cleaner does not present a fire hazard and is much cheaper than the other type of cleaner.

A good, soft soap solution may be used for much of the routine cleaning of painted and unpainted airplane surfaces. An excellent method of preparing a soft soap solution is to fill a 50-gal. drum within a few inches of the top with clean, soft water. A porous bag, such as a burlap bag or a cheesecloth bag, is filled with 50 to 75 lb. of soft soap and allowed to hang in the water, suspended from a stick placed across the top of the drum. A comparatively thin bag will hold the soap if the bag is placed in the water and the soap added without removing the bag from the water. The soap will gradually dissolve and will not settle to the bottom, taking a long time to dissolve.

Approximately 1 qt. of this concentrated soap solution may be added to 5 gal. of water. This diluted solution may be applied with rags, mops, or sponges to the parts being cleaned. After the part is cleaned, it should be thoroughly rinsed with clear water. Grease spots, exhaust stains, and other spots that are hard to remove may be treated with liquid cleaner or polished with a scouring powder, or even dope thinner may be used. Dope thinner, of course, cannot be used on anything but unfinished metal surfaces. Unleaded gasoline, alcohol, kerosene, carbon tetrachloride, Varsol, or other solvents may be used. Kerosene will burn if ignited, and kerosene vapor may catch fire readily if exposed to an open flame or sparks. Alcohol and gasoline are highly inflammable, and great care should be taken if these compounds are used for cleaning purposes. Carbon tetrachloride is noninflammable. Naphtha and carbon tetrachloride are sometimes mixed in equal parts and used as a cleaner. When using any of the inflammable cleaning compounds, fire extinguishing equipment must be available and in charge of a person entirely familiar with its use.

Carbon tetrachloride or solutions containing this compound should not be applied to heated parts because it will give off poisonous fumes when heated. Breathing the vapors of the cleaning solution should be avoided, and they should not be allowed to come in contact with the skin. Both the soap solution and the solvent may be applied by means of pressure nozzles or spray guns. The spray is more often used for gasoline, kerosene, or white furnace oil than for the water-solvent cleaners. Carbon tetrachloride should not be used in the form of a spray. The spraying of the inflammable cleaners should never be done except in the open air. The person using the spray and other personnel should always keep to the windward side to avoid breathing the fumes.

Salt Water. If the aircraft has been exposed to salt water spray, all traces of the salt water should be removed by thoroughly rinsing all the exposed parts of the aircraft with fresh water at the end of each day's flying. A light coat of rust-preventive compound should then be applied to all exposed fittings on which corrosion is likely to occur.

GENERAL MAINTENANCE OF THE AIRPLANE

These are such parts as exposed rivets and bolts, control cables, control-surface hinges, exposed parts of the landing gear, retracting pistons, and other exposed parts which are not protected permanently from corrosion. The manufacturer's recommendations should be followed for all oils and lubricants.

VI THE LIGHT AIRPLANE

The greater number of airplane mechanics are engaged in the light airplane field. The following section will give briefly erection procedures, general maintenance instructions, installation of the major component parts, and maintenance, repair, and general service inspections of a light composite type of airplane.

It would be impossible to give detailed descriptions of all the light aircraft being manufactured today. Most light aircraft, however, are of composite construction. They are, in general, made of metal, wood, and fabric construction. The general principles given for the light airplane may be applied to any type of airplane by following the manufacturer's manuals.

Uncrating and Preparing for Erection. The light airplane is usually shipped in a crate, disassembled into its major component parts or assemblies. These parts consist of the wings, the engine and its mount, the fuselage, the tail group, the landing gear, and their various attachments. The fuselage usually occupies the main part of the crate, with the wings attached one on each side of the crate.

Care should be taken in removing the side or end of the crate to avoid damage to any part which may be attached to it. One wing is usually removed first, and then the fuselage is removed. It is sometimes necessary to remove other parts before the wings and fuselage are accessible. Racks should be available for the wings and tail parts, and proper supports for the fuselage.

The front part of the fuselage is usually mounted on one horse with another horse under the tail. The parts should not be placed on the ground or floor, as this may result in damage. The parts are usually tagged with their proper part number and marked right and left where necessary. Each tag usually has on it the serial number of the airplane. All parts should be assembled in their proper order. In most light airplanes, it is necessary that the fuselage should be weighted at the tail

THE LIGHT AIRPLANE

to prevent nosing over after the main landing-gear assembly is in place. A weight of from 30 to 50 lb. is sufficient.

The corrosion-resistant materials should be carefully removed from all exposed metal parts. This material may damage the fabric finish and should be removed immediately by using gasoline if, by chance, any of it is smeared onto the fabric surface. The cowling should be removed from the engine, and the engine thoroughly cleaned of the anticorrosion material. The moistureproof wrappings, silica gel, and de-

Fig. 48. A light airplane of composite construction. (Courtesy Taylorcraft Aviation Corporation)

hydrating plugs should be removed. The engine should be thoroughly drained of any corrosion-resistant material. Care should be taken not to tear the bags of silica gel. If any of this material gets into any part of the engine, that part of the engine must be disassembled and thoroughly cleaned. After excess corrosion-preventive material has been removed from the engine and the spark plugs, the spark plugs should be installed "finger tight." All parts should be thoroughly inspected before erection.

The fuselage should be inspected for breaks or tears in the covering, damaged fittings, broken plastic in the windshield or windows, and any wrinkling of the fuselage covering which would indicate distortion or twisting of the main fuselage structure. The wings should be inspected for cuts, tears, breaks, damaged fittings, and to see that all vent holes are open. Any wrinkles in the wing fabric might indicate broken or distorted internal parts. The tail surfaces, landing gear, and all other disassembled parts should be carefully examined for visible signs of damage. Wrinkling of the covering should lead to a thorough inspection of the inner structure, either through regular inspection openings or by cutting and opening the fabric at the wrinkled place.

Erection Procedure. With the fuselage properly mounted on horses, the landing gear is attached. Care should be taken that the fuselage rests on such of its parts as will give it firm support without damage to the covering or to weak internal parts of the structure. The landing-gear V's are attached to the fuselage by proper bolts. Be sure that all bolts are inserted from the proper side of the fitting. The brake cables then are attached and threaded through the proper fair-leads into the fuselage. The tail-wheel assembly is next attached. It is absolutely essential that, after the bolts, nuts, and other fastenings are in place, they be properly safetied.

If the aileron push-pull tubes or cables have been removed, the proper attachments should be installed in the fuselage before attaching the wings. The wings are next attached to the fuselage in the manner directed by the manufacturer. The fuel lines are then connected to the wing tanks. Navigation-light wires are properly connected, and all fairings properly installed.

If the airplane is equipped with a generator, the generator wires should be connected. The tail should then be assembled in a manner directed by the manufacturer. Be sure that all fastenings are properly safetied and that the movable surfaces are free and do not bind when moved. The engine and its mount are next installed, and all wiring, fuel lines, oil lines, control cables, or control rods or wires are installed and checked.

Rigging. The engine should be carefully inspected for any damage which might have occurred in transit. With the spark plugs removed, the engine should be rotated through several revolutions by hand to determine that the cylinders and valves operate freely and that there is no accumulation of corrosion-resistant material in the cylinders. If the valves have any tendency to stick, the stems may be lubricated with a mixture of gasoline and lubricating oil. The engine should then be rotated by hand until the valves are operating properly. Never start an engine hoping that parts will adjust themselves and begin operating after the engine has started. Any necessary repairs or adjustment should be made before attempting to start the engine.

In most light airplanes, if all parts have been properly installed, additional rigging is not necessary. The wings are usually held in their

THE LIGHT AIRPLANE

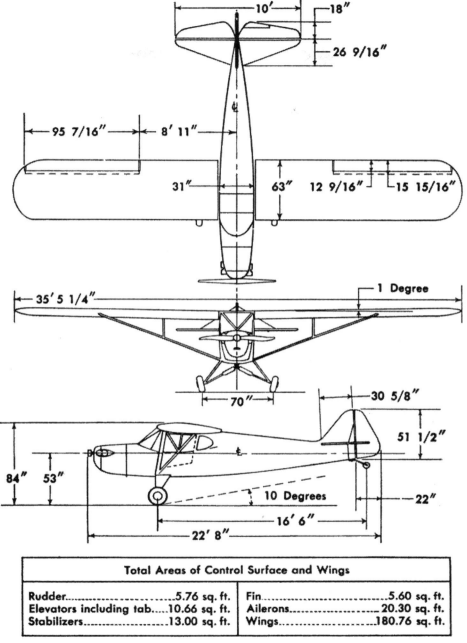

Total Areas of Control Surface and Wings	
Rudder..................................5.76 sq. ft.	Fin.......................................5.60 sq. ft.
Elevators including tab.....10.66 sq. ft.	Ailerons............................20.30 sq. ft.
Stabilizers........................13.00 sq. ft.	Wings.............................180.76 sq. ft.

Fig. 49. Three-view diagram of light aircraft. (Courtesy Taylorcraft Aviation Corporation)

AIRCRAFT MAINTENANCE AND SERVICE

proper position by the fixed length of the supporting struts, and the tail surfaces are held by their rigid tie rods. If it is necessary to rerig the airplane, the following is given as a typical procedure.

A spirit level is placed on the part of the wing designated by the manufacturer to rig the wing properly. This is usually on the lower side of the first full rib inboard from the tip of the wing. Adjustment

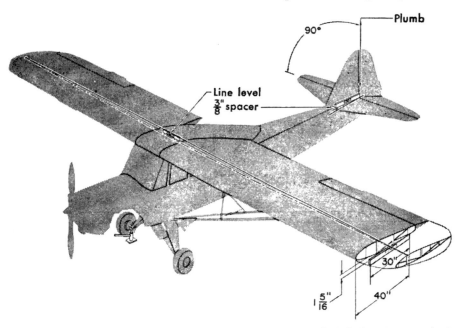

Fig. 50. Rigging and leveling diagram. (Courtesy Taylorcraft Aviation Corporation)

is usually made by the adjusting arrangement on the rear strut. The top attaching bolt is removed, and the adjusting barrel is screwed in or out until the proper position has been attained.

In rigging the tail group, a spirit level is usually placed on the trailing edge of the fin and stabilizer. The tail should be rigged plumb and level by adjusting the tie rods. Before this adjustment is undertaken, it is necessary that the fuselage be level laterally and that the airplane is placed in a level flying position. Brakes and control surfaces are checked to see that they are in proper adjustment, and all fastenings are safetied. After the proper ground run-in of the engine, the oil should be drained and fresh oil installed.

With the proper amount of fuel in the tanks, the airplane should be given its check flight. If the airplane is properly rigged, it should fly straight and level "hands off" under full-load conditions with the ele-

THE LIGHT AIRPLANE

vator trim tabs at neutral. If the airplane flies wing heavy, this condition may be corrected by washing in the heavy wing or washing out the light wing. It is recommended that the condition be corrected, where possible, by washing out the light wing. A wing is washed in by shortening the rear strut. A half turn at a time is usually advisable. If the wing is to be washed out, the strut should be lengthened not more than a half turn at a time. If the airplane flies nose heavy, both wings should be washed in for correction. If the airplane flies tail heavy, both wings should be washed out for correction. If the airplane tends to yaw right or left, the trim tab on the rudder should be used to correct this condition.

Handling and General Maintenance. With light airplanes, special handling equipment is not usually necessary to move this type of airplane on the ground. They may be pushed or pulled by the wing strut close to the fuselage or close to the wing-attachment fittings. Never push or pull on the center of the struts. Do not attempt to move the airplane by pulling or pushing on the propeller blades.

Usually this type of airplane should not be moved backwards without lifting the tail. The tail should only be lifted by the indicated lift points. The tail should never be lifted, pushed, or pulled by taking hold of the stabilizer, tail surfaces, or tie rods. Most airplanes may be lifted from the ground by following the manufacturer's hoisting directions.

In jacking up the airplane, the jack should be placed at the points designated by the manufacturer. To level the airplane laterally, a cord should be stretched from wing tip to wing tip over the front spar. A level should be placed on the center of the cord, and the airplane lifted by jacking up the lower wheel. In tying down an airplane, the ropes should be attached as directed by the manufacturer.

Ground Operating Instructions. In entering an airplane, the proper steps or wing walks should be used. Do not step on any parts, such as the strut or wing, unless marked specifically to be used as a step. All external locks on the control surfaces should be removed before entering the cockpit to start the engine.

Engine Starting. The following check should be made as soon as the pilot enters the pilot's compartment:
1. Ignition switch in the OFF position
2. Fuel shut-off valve in the ON position
3. Parking brake ON

AIRCRAFT MAINTENANCE AND SERVICE

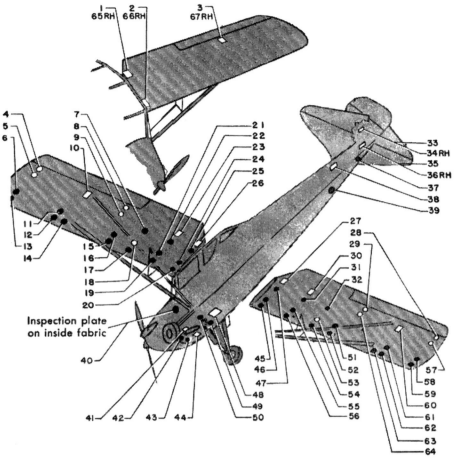

Fig. 51. Access and inspection provisions.

Key to Fig. 51. Access and inspection provisions:

(1) aileron control pulley and push-pull rod, (2) wing-tank vent line and wing electrical wiring, (3) aileron control pulley and drag link, (4) drag strut, drag wires, and aileron hinge bracket, (5) drag strut, drag wires, and aileron hinge bracket, (6) drag strut and drag wires, (7) drag strut and drag wires, (8) drag strut, drag wires, and aileron hinge bracket, (9) drag strut, drag wires, and aileron hinge bracket, (10) aileron control pulley and drag link, (11) drag strut and drag wires, (12) drag strut and drag wires, (13) drag strut and drag wires, (14) lift strut fitting, (15) drag strut and drag wires, (16) drag strut and drag wires, (17) drag strut, drag wires, and jury strut fitting, (18) drag strut, drag wires, and jury strut fitting, (19) drag strut and drag wires, (20) drag strut and drag wires, (21) drag strut and drag wires, (22) drag strut and drag wires, (23) aileron control cables and turnbuckles, (24) drag strut and drag wires, (25) drag strut and drag wires, (26) aileron control pulley and push-pull rod, (27) aileron control pulley and push-pull rod, (28) drag strut, drag wires, and aileron hinge bracket, (29) drag strut, drag wires, and aileron hinge bracket, (30) aileron control cables and turnbuckles, (31) drag strut and drag wires, (32) drag strut and drag wires, (33) upper elevator horn, control cable, and turnbuckle, (34 RH) upper elevator horn control, (35) lower elevator horn control, (36 RH) lower elevator horn, control cable and turnbuckle, (37) tail wheel (wrench hole), (38) rudder control cables, (39) rudder control pulleys, (40) crank-type trim tab,

THE LIGHT AIRPLANE

(41) brake control cables and pulleys, (42) rudder and brake pedals, (43) rudder and brake pedals, (44) brake control cables and pulleys, (45) drag strut and drag wires, (46) drag strut and drag wires, (47) drag strut and drag wires, (48) rear seat, (49) rudder and brake pedals, (50) rudder and brake pedals, (51) drag strut and drag wires, (52) drag strut and drag wires, (53) drag strut, drag wires, and jury strut fitting, (54) drag strut, drag wires, and jury strut fitting, (55) drag strut and drag wires, (56) drag strut and drag wires, (57) drag strut, drag wires, and aileron hinge bracket, (58) drag strut and drag wires, (59) drag strut and drag wires, (60) aileron control pulley and drag link, (61) drag strut and drag wires, (62) drag strut and drag wires, (63) lift strut fitting, (64) drag strut, drag wires, and aileron hinge bracket, (65 RH) aileron control pulley and push-pull rod, (66 RH) wing-tank vent line and wing electrical wiring, (67 RH) aileron control pulley and drag link. NOTE: The access and inspection provisions illustrated in white represent holes over which cover plates are fastened, those illustrated by shading represent provision for holes by application of reinforcements doped to the fabric cover. (Courtesy Taylorcraft Aviation Corporation.)

Fig. 52. Hoisting diagram. (Courtesy Taylorcraft Aviation Corporation)

Fig. 53. Tie down diagram. (Courtesy Taylorcraft Aviation Corporation)

AIRCRAFT MAINTENANCE AND SERVICE

4. Controls should move freely
5. Throttle should operate freely throughout its full range.

To start a cold engine, the following should be performed in the order given:

1. Ignition switch in the OFF position
2. Fuel shut-off valve in the ON position
3. Carburetor heat control in the full-cold position
4. Mixture control in the full-rich position
5. The throttle should be closed
6. The engine should be primed with two or three strokes on the priming pump and turned over one or two complete revolutions by turning the propeller. Overpriming may wash the lubricating oils from the cylinder walls
7. The ignition switch should be turned to BOTH. This turns on both magnetos
8. With the propeller in the proper position, the switch should be turned on and the propeller pulled through rapidly with as much snap as possible
9. If the engine fails to start, the above procedure should be repeated. If the engine appears to be flooded, it should be turned backward by means of the propeller through several revolutions with the throttle wide open and the switch OFF. The throttle should then be closed and the starting procedure repeated without further priming.

When starting a warm engine, no priming is usually necessary. Try starting the engine without either turning over or priming. If it does not start immediately, proceed as follows. With the ignition in the OFF position, turn the engine over two or three times in the direction of normal rotation and try again. The primer should only be used if the engine then fails to start. As soon as the engine starts the oil-pressure gauge should be checked and, if the oil pressure does not show within 30 sec. after the engine starts, the engine should be stopped and the trouble located and remedied.

If the engine should backfire in starting and cause a fire in the carburetor before the engine starts, a hand fire extinguisher should be used to extinguish the blaze. If the fire occurs in the carburetor after the engine is started, the throttle should be opened immediately. This causes a fire to be sucked into the engine where it can do no harm. Never cut the switch to extinguish a carburetor fire, if the engine is already running.

THE LIGHT AIRPLANE

Major Components. The wings of most composite light aircraft are fabric covered. The spars are of wood, and the ribs may be of wood or stamped from metal alloy. The wooden ribs are nailed to the spars, while the metal ribs are usually fastened to the spars by means of metal clips. The leading edge may be of plywood or of light, aluminum alloy sheet. A wooden leading edge is glued and nailed to the ribs and a reinforcing strip on the spar, while a metal leading edge is usually attached by means of sheet-metal screws.

Most light-airplane wings of this type have tubular steel wing bows. The drag struts are usually of tubular steel and held in place by steel drag and antidrag wires. Fuel tanks are usually installed near the butt end of the wing panel or in a center section, or may be installed under the cowl in front of the pilot. The ailerons resemble the wings in their construction. Most ailerons of this type have wooden spars and may have either aluminum alloy ribs or wooden ribs. Ailerons are usually fabric covered. The wings on a high-wing type of airplane are supported by means of streamlined, tubular, lift and jury struts.

Installing a Wing Panel. To install a wing panel, the wing should be lifted to its proper position and supported at points close to the lift and jury strut fittings. Do not support the wing on the ribs or fabric. The fuselage fittings are fastened into place and all bolts are safetied immediately after installation. The fuel lines, light wires, and aileron controls are then attached. The lift strut and the jury strut are attached at their proper fittings. All bolts should be safetied immediately on installation. The ailerons, if they are not already installed, may be lifted into position and attached at the hinge fittings. The controls are then attached to the ailerons and safetied.

Removing a Wing Panel. To remove a wing panel, the fuel tank should be drained and the wing-butt fairing removed. The fuel lines are then disconnected, as are the light wires and aileron controls. The lift and jury struts are disconnected at their upper ends by removing their fastening. The wing must be carefully supported at the strut fitting.

Before detaching the wing panel from the fuselage, provision should be made for its careful handling after removal. Dropping the wing may cause serious damage. To remove the ailerons from the wing panel, the controls should be disconnected and the pins removed from the hinges. This will usually allow the aileron to be lifted free. When removing the lift or jury struts, the upper fitting should first be disconnected and the wing properly supported. Do not disconnect the lower end of the struts first.

Most manufacturers recommend the repair of only minor damage to the fabric. No attempt should be made to repair damaged lift or

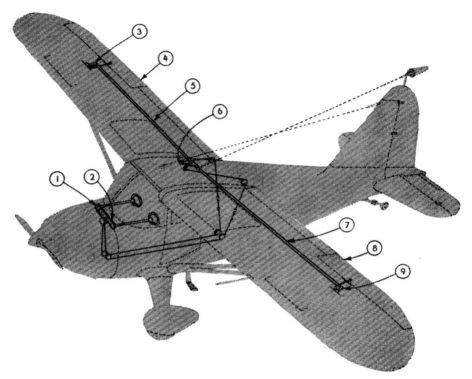

Fig. 54. The aileron control system showing (1) sprocket, (2) chain, (3) bell crank, (4) right aileron, (5) push-pull tube, (6) bell crank, (7) push-pull tube, (8) left aileron, (9) bell crank. (Courtesy Consolidated Vultee Aircraft Corporation, Stinson Division)

jury struts. Aileron control-pulley brackets, hinge brackets, and hinge and control-point fittings must be replaced if in need of repairs. The adjusting plugs and the struts must also be replaced if damaged.

At overhaul periods, the wing fairing should be removed. Wing attachments should be inspected, and all bolts tightened, if necessary, and resafetied. All inspection openings should be opened and the inner structure examined. If inspection openings are not available, the fabric should be cut in order to examine the internal structure. All visible bolts and nuts should be examined and, if necessary, tightened and resafetied. All drag and antidrag wires must be tight. If any looseness

THE LIGHT AIRPLANE

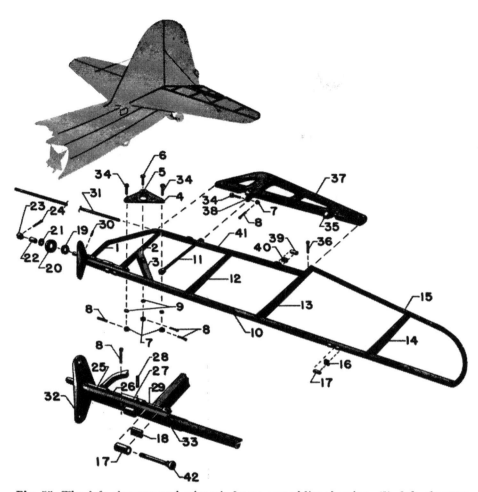

Fig. 55. The left elevator and trim-tab frame assemblies showing (1) left elevator butt trailing edge, (2) left elevator No. one rib assembly, (3) elevator tab bellcrank mount assembly, (4) elevator trim-tab bell crank, (5) disk, (6) aircraft bolt, (7) shear nut, (8) cotter pin, (9) washer, (10) elevator leading edge, (11) elevator tab link assembly, (12) left elevator No. 2 rib, (13) left elevator No. 3 rib, (14) left elevator No. 4 outer rib, (15) left elevator trailing edge, (16) filler block, (17) sleeve, (18) filler block, (19) pulley spacer washer, (20) elevator tab control rear pulley, (21) lock washer, (22) trim-tab control spacer, (23) shear nut, (24) cotter pin, (25) trim-tab control sleeve, (26) elevator tab drive screw, (27) elevator tab connector, (28) elevator trim-tab bearing pin, (29) elevator tab drive assembly, (30) trim-tab drive pin, (31) elevator tab bearing pin, (32) elevator horn, (33) elevator horn reinforcement sleeve, (34) aircraft bolt, (35) elevator tab hinge assembly, (36) cotter pin, (37) elevator trim-tab frame, (38) elevator tab hinge and horn assembly, (39) hinge tube, (40) filler block, (41) left elevator cutout tube, and (42) tail surface hinge pin. (Courtesy Taylorcraft Aviation Corporation)

of drag or antidrag wires is discovered, a complete check of all the wires and the spars is recommended. All connecting links and pulleys in the aileron control system should be lubricated, and any excessive play should be removed. Aileron hinge pins, bushings, and fittings that show excessive wear should be replaced. The attaching bolts of the lift and jury struts should be loosened and, if any movement is possible between the adjoining parts or the holes have become elongated, the fittings must be replaced. If no play is discovered, the bolt should be retightened and resafetied.

Tail Group. The tail surfaces are usually built of welded tubular steel. The ribs may be made of formed metal or of wood. The tail surfaces are usually fabric covered. The fixed tail surfaces are generally bolted to the fuselage and braced with tie rods. The tie rods are usually attached to self-aligning fittings. On the rudder and elevator are hinge fittings. The rudder and elevator are fastened into place with hinge pins which pass through corresponding fittings on the fixed surfaces. The elevator trim tab is often constructed of a single piece of plywood and is fabric covered.

Various parts of the tail group may be disassembled in accordance with the manufacturer's directions. The repair of minor fabric damage is the only maintenance repair recommended by most manufacturers for the tail surfaces. All fastenings and tie rods should be tight and safetied. The tail surfaces, tie rods, tie-rod fittings, hinge pins, bushings, and rudder and elevator trim tabs may be replaced. All bolt hinge pins and bushings should be examined at frequent intervals to make sure that they are firmly fastened and safetied. Any of these parts that show excessive wear should be replaced immediately.

Fuselage. The fuselage of most composite-type light aircraft is constructed of welded steel tubing and is fabric covered. The fuselage structure should be checked at regular intervals for damage to the main structure or wooden fairing parts. Minor damage to the fabric covering should be repaired at once. Windshields, windows, and any other transparencies should be kept in good condition. The fastenings should be secured, and any damaged parts should be replaced with a new part. All bolts and pins and all fittings should be examined frequently and, if necessary, tightened and resafetied. Any pins, fittings, or fastenings that show excessive wear should be replaced. The vertical fin, which is usually of separated construction, is joined to the fuselage by bolts and fittings. The fin is usually covered with fabric which is integral

AIRCRAFT MAINTENANCE AND SERVICE

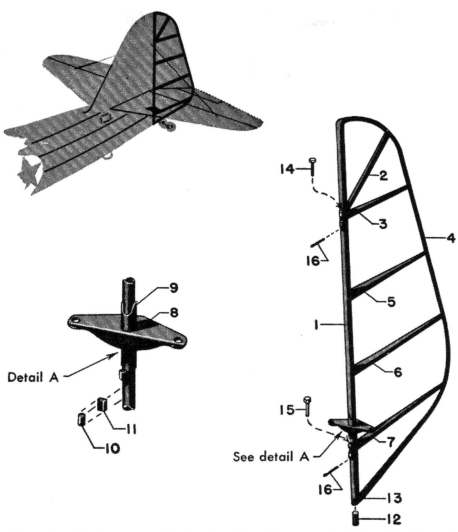

Fig. 56. A rudder frame assembly showing (1) rudder post tube, (2) rudder diagonal brace, (3) rudder No. 1 upper rib, (4) rudder trailing-edge tube, (5) Rudder No. 2 upper middle rib, (6) rudder No. 3 lower middle rib, (7) rudder No. 4 lower rib, (8) rudder control lever, (9) rudder control lever sleeve, (10) hinge sleeve, (11) filler block, (12) cork, (13) reinforcement strap, (14) tail surface hinge pin, (15) rudder lower hinge pin, (16) cotter pin. (Courtesy Taylorcraft Aviation (Corporation)

AIRCRAFT MAINTENANCE AND SERVICE

with the fabric covering of the fuselage. To remove the fin, it is necessary to cut the fabric in order to get at the bolts and fittings.

Landing Gear. All landing-gear parts should be examined carefully at frequent intervals to determine whether any damage has occurred. Tires should be kept properly inflated, and the brake assembly in perfect condition. The main landing wheels may usually be removed by taking off the hub cap, axle nut, and the keyed washer from the axle. The wheel will then come free. Most wheels are made in two parts,

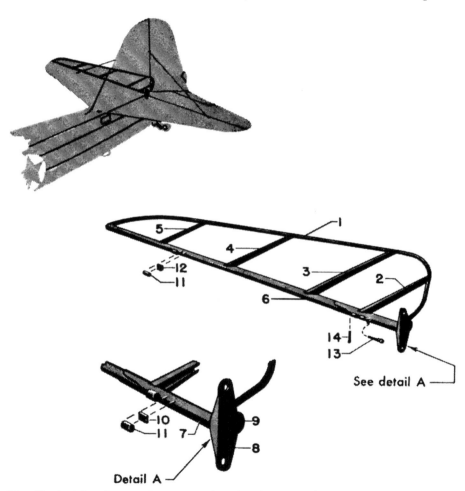

Fig. 57. A right elevator frame assembly showing (1) elevator trailing-edge tube, (2) elevator No. 1 inner rib, (3) elevator No. 2 rib, (4) elevator No. 3 rib, (5) elevator No. 4 outer rib, (6) elevator leading edge, (7) elevator-horn reinforcing sleeve, (8) elevator horn, (9) right elevator-horn reinforcement, (10) filler block, (11) sleeve, (12) elevator hinge filler block, (13) tail-surface hinge pin, and (14) cotter pin. (Courtesy Taylorcraft Aviation Corporation)

THE LIGHT AIRPLANE

fastened together by bolts. Tires are removed by dismantling the wheel. Tire-dismounting tools should not be used.

Shock-cord boots should be removed for inspection purposes. Unless it is necessary to replace shock cord in an emergency, disassembly of the tie strut, except at a repair depot, is not recommended. No attempt should be made to repair a wheel. Most wheels are made from stamped

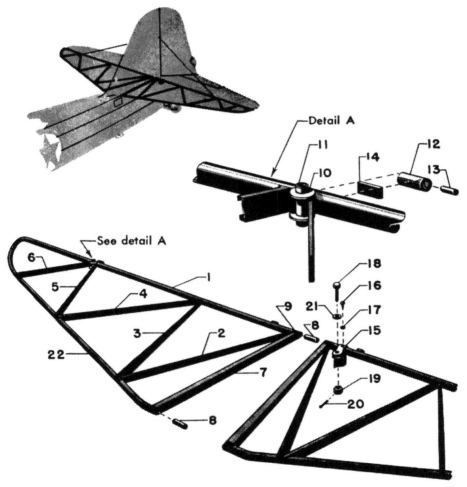

Fig. 58. A stabilizer frame assembly showing (1) stabilizer trailing-edge tube, (2) No. 2 stabilizer rib, (3) No. 3 stabilizer rib, (4) No. 4 stabilizer rib, (5) No. 5 stabilizer rib, (6) No. 6 stabilizer rib, (7) No. 1 inner stabilizer rib, (8) stabilizer cork, (9) blank washer, (10) reinforcement wire attaching washer, (11) tie-rod bolt sleeve, (12) aileron hinge bearing tube, (13) control bushing, (14) filler block, (15) clamp assembly, (16) screw, (17) lock washer, (18) aircraft bolt, (19) shear nut, (20) cotter pin, (21) washer, and (22) stabilizer leading-edge tube. (Courtesy Taylorcraft Aviation Corporation)

AIRCRAFT MAINTENANCE AND SERVICE

and forged steel and are heat-treated. Bushing bolts and nuts may be replaced in any part of the landing gear. Bent axles or other bracings should be replaced if damaged. A brake lining may be replaced by a qualified person if the necessary equipment is available. The brake lining should be replaced in accordance with the manufacturer's recommendations.

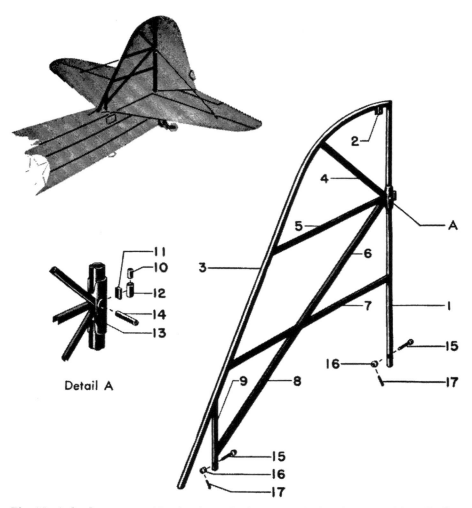

Fig. 59. A fin frame assembly showing (1) fin post, (2) bracket assembly, (3) fin leading-edge tube, (4) fin upper diagonal bracket, (5) fin top rib, (6) fin center diagonal brace, (7) fin bottom rib, (8) fin lower diagonal brace, (9) fin front vertical tube, (10) control bushing, (11) filler block, (12) aileron hinge bearing tube, (13) guy wire reinforcement sleeve, (14) tie-rod bolt sleeve, (15) aircraft bolt, (16) shear nut, and (17) cotter pin. (Courtesy Taylorcraft Aviation Corporation)

THE LIGHT AIRPLANE

Any damaged part of the cowl may be replaced. It is important that all cowl fastenings be tight and free from excessive wear. Large dents or tears in the cowling should be repaired immediately in accordance with approved methods. A cracked, torn, or severely damaged piece of cowling should be replaced.

Fuel System. The fuel system on most light airplanes is of the gravity-flow type. Most light airplanes carry from 12 to 16 gal. of fuel.

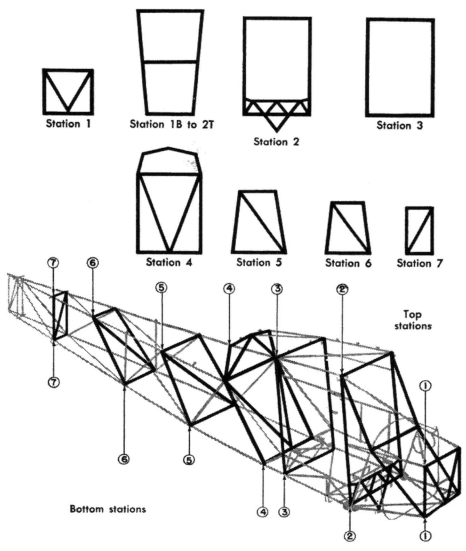

Fig. 60. Stations and frame diagram. (Courtesy Taylorcraft Aviation Corporation)

AIRCRAFT MAINTENANCE AND SERVICE

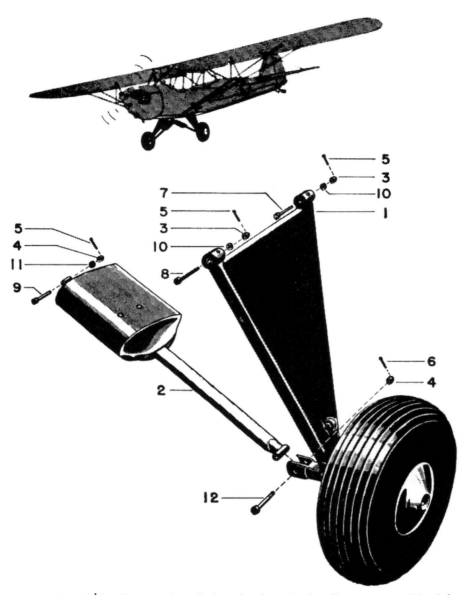

Fig. 61. A main landing-gear installation showing (1) Landing-gear assembly, left, (2) aircraft bolt, (3) aircraft bolt, (4) washer, cadmium-plated, (5) castle nut, (6) cotter pin, (7) (landing gear tie) strut assembly, (8) aircraft bolt, (9) aircraft bolt, (10) plain washer, (11) castle nut, (12) cotter pin. (Courtesy Taylorcraft Aviation Corporation)

THE LIGHT AIRPLANE

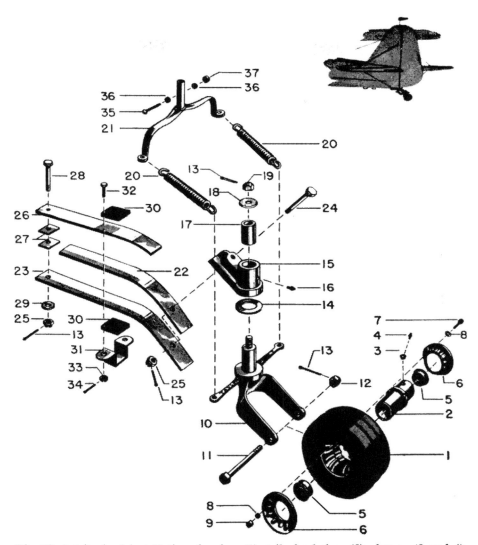

Fig. 62. A tail-wheel installation showing (1) tail-wheel tire, (2) sleeve, (3 and 4) fittings, (5) bearing, (6) disk, (7) bolt, (8) washer, (9) nut, (10) tail-wheel fork, (11) aircraft bolt, (12) shear nut, (13) cotter pin, (14) brass washer, (15) block, (16) grease fitting, (17) brass bushing, (18) steel washer, (19) castle nut, (20) rudder steering coil spring, (21) rudder arm, (22) center spring leaf, (23) bottom spring leaf, (24) aircraft bolt, (25) castle nut, (26) top spring leaf, (27) tail-wheel spring spacer, (28) aircraft bolt, (29) washer, (30) tail-wheel spring pad, (31) lower spring clip, (32) aircraft bolt, (33) shear nut, (34) cotter pin, (35) roundhead machine screw, (36) plain washer, (37) self-locking nut. (Courtesy Taylorcraft Aviation Corporation)

AIRCRAFT MAINTENANCE AND SERVICE

Fuel tanks are usually located either under the cowl in front of the pilot, in the root end of the wing, or in the center section. Usually there are at least two fuel tanks. In one light airplane there is a fuel tank in each wing with a capacity of 6 gal. and a collector tank in the front of the fuselage holding 2 gal. All tanks are interconnected and vented. In one particular airplane, all tanks may be filled from either wing tank.

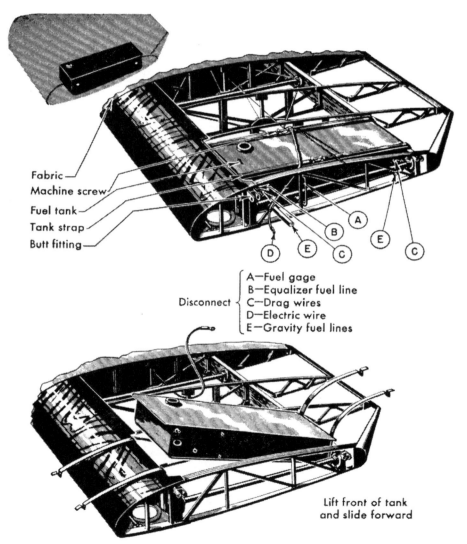

Fig. 63. Removal of fuel tank from wing. (Courtesy Taylorcraft Aviation Corporation)

Fuel may be taken on through either the right or left wing tank. If an airplane is standing on uneven ground with one wing low, the low wing tank may fill up first. In fueling, the low tank should be filled first and after the filler cap has been put in place, the high wing tank may be filled. On most airplanes, fuel may be used from either tank by operation of the proper valves.

Screens are usually located at the take-offs of each tank. At this point, there is usually a sump which collects water or other impurities from the fuel. Before removing any part of the fuel system, the entire fuel system should be drained by opening the drain cock. If the airplane has been standing inactive for a considerable length of time, the drain cock should be opened to be sure that there is no collection of water in the system. To remove fuel tanks, it is usually necessary to remove the fabric from that part of the wing in which the tank is located. The fuel filter bowl may usually be removed after placing the fuel shut-off control in the OFF position. The fuel filter bowl should then be removed, allowing the fuel to drain from that portion of the system.

Tanks may usually be replaced as units, and all fuel lines may be replaced. No repairs should be made on fuel lines, but fuel lines may be straightened if only slightly bent. If leaks develop in any of the shut-offs, these parts should be replaced.

Instruments. All repairs to instruments should be made by a thoroughly qualified instrument repairman. It is possible, however, to remove and replace faulty instruments. Any instruments removed should be sent to an instrument overhaul station or the manufacturer. Adjustments should not be attempted on instruments, and each instrument should be replaced as a unit.

Control System. On most light airplanes, the control system consists of either side-by-side or front-and-rear interconnected rudder pedals. These pedals are connected with cables running approximately along the two lower longerons.

The aileron system usually consists of either two control sticks or interconnected control wheels. The cables usually run up along each side of the cabin, and the ailerons are interconnected. Each wing usually has a closed cable system running between pulleys located at the root of the wing and at the aileron control point. The controls may either be of cable or of the push-pull tube type. Some systems have cables controlled from the cabin to the wing with a push-pull tube arrangement from the stick or wheel control.

AIRCRAFT MAINTENANCE AND SERVICE

The elevator control system is usually connected to the stick or wheel and is operated by cables connecting the elevator horn to the torque tubes connected with the stick or wheel. Most airplanes of this type have an elevator trim tab which is operated by a crank or wheel located in the cabin. The cables may be repaired, if necessary, in ac-

Fig. 64. Inspecting the instruments on a light aircraft. (Courtesy Piper Aircraft Corporation)

cordance with approved procedures, but no structural repairs are permitted to the control sticks, wheels, or torque tubes. Any part of the control-surface system may be replaced. Adjustments to the control systems are usually made by means of turnbuckles which should always be properly safetied after adjustment. Any adjustment to the cables should not tighten them enough to cause binding at any point in their normal travel.

All cables should be inspected at regular intervals for indications of fraying. This is apt to take place where the cables run over pulleys or through fair-leads. All pulleys should be carefully checked for freedom of movement and for checks, splits, or worn places. Sticking pulleys should be carefully lubricated and freed. Pulleys should not be excessively lubricated as this will cause collection of dust and dirt which might interfere with the proper operation. All damaged pulleys should

THE LIGHT AIRPLANE

be replaced. Any indication of elongated holes or excessive wear in the control system is reason for replacing these worn parts.

Finishing after Repairs. After repairs are made or new parts installed, they should be finished in a manner which compares favorably with high-grade commercial practice or the minimum standards set up

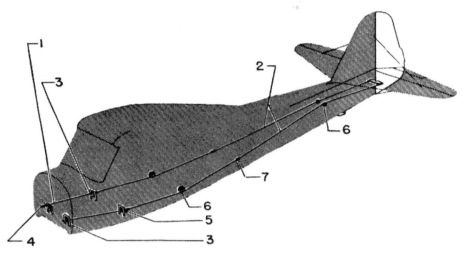

Fig. 65. A rudder control system showing (1) rudder pedal to rudder pedal cable assembly, (2) rudder pedal to horn cable assembly, (3) rudder pedal assembly, (4) rudder pedal spring, (5) cable adjustment plate, (6) small pulley, (7) cable guide bushing. (Courtesy Taylorcraft Aviation Corporation)

by the Civil Aeronautics Administration. All replacement parts should be of standard manufacture, and any construction in the form of a repair or replacement should be approved by an inspector of the Civil Aeronautics Administration.

All metal parts on the airplane should be covered with a protective coating, either by plating or painting. All metal parts of the airplane subject to corrosion, which are not painted, must have a corrosion-resistant material applied.

The fuselage structure tubes should be coated with zinc chromate primer. This may be applied by either a spray or a brush. Any exposed surfaces of the airplane structure should also receive finish coats of lacquer or dope. All wooden parts of the structure should be treated with a proper wood sealer. This sealer should be applied after assembly. Care should be taken that the wood sealer is not applied to a surface which is to be glued. Exposed portions may be finished with the proper color of lacquer or dope. The fabric should be treated with a series of applications of cellulose dope, which should be applied at temperatures between 70° F. and 90° F. Dope should always be applied in a space free from dust and strong drafts. Care should be taken in applying the fabric covering to see that the new fabric is not sufficiently tight to cause distortion of the wooden parts when it shrinks and tightens with the application of the dope.

Doping and finishing should be done in accordance with approved methods as recommended by the manufacturer or the Civil Aeronautics Administration.

Service Inspection. This section is planned to lay before the maintenance crew the minimum requirements for periodic inspection, cleaning, servicing, and lubrication of the average light airplane and its component parts. These procedures and inspections should be followed by all maintenance personnel. If carefully followed, the aircraft should be maintained in good operating condition at all times. All inspections must be made by properly qualified personnel.

Preflight Inspection. Before making a preflight inspection, the member of the maintenance crew making the inspection should check the log books or flight reports of the airplane to see whether any notations have been made by the pilot or flight crew. An inspection form should always be used and carefully followed.

Fuel System. All fuel tanks and lines should be checked for any evidence of leaking and for the quantity of fuel present. Check the fuel level in each tank and check the fuel gauges to see that they are indicating the correct amount of fuel in the tank. The filter bowls should be examined and any dirt, water, or bubbles should be removed. Fuel shut-off valves should be in the closed position. The filter bowl should be removed and cleaned. After replacing the filter bowl, it should not be tightened into place until after the fuel has been turned on and the bowl has become completely filled with fuel. The bowl should never be tightened in place if there are air bubbles still present. After the bowl has been tightened, it must be safetied with safety wire.

Any drain valves in the system should be opened and allowed to drain for one or two seconds. The valves should then be closed and safetied. Check to see that the proper grade of fuel is in the tanks. If the necessary grade is not at hand, the next higher octane rating of fuel may be used. Fuel-tank filler caps should be replaced with the vent pointing forward. If the cap is installed with the vent pointing backward, a suction may be set up which will prevent the proper flow of fuel.

Oil System. The quantity of oil should be checked and, if necessary, the oil tank should be filled to the proper level with the grade of oil recommended by the manufacturer. All oil lines should be examined for evidence of leaking. The by-pass relief valve should be checked for proper functioning. If the atmospheric temperatures are low, it may be necessary to change the oil to a lighter grade. A heavier grade may be necessary if the atmospheric temperatures are high.

Propeller. Before inspecting a propeller, it is always necessary to be sure that the ignition switch is in the OFF position. Even then, a person should always stand clear of the propeller to prevent injury in case the engine should accidentally start because of a loose magneto ground wire or some other defect in the wiring system. The propeller should be carefully examined for cracks, dents, and splits, and to see that it is tight on the hub. If a wooden propeller is loose on the hub, the vibration will cause the propeller to crack. All safetying devices should be checked.

Battery. If the airplane is equipped with a storage battery, it should be checked for leaks, loose or corroded terminals, and broken cables. The battery leads should be examined to see that the insulation is in good condition. If any leaks are indicated, all parts of the airplane structure which may have come in contact with the battery acid should be carefully examined. Battery acid should be washed from all parts of the airplane by using a solution of sodium bicarbonate, which is common baking soda. About 1 tablespoonful of the soda should be placed in each pint of water. After the fizzing has stopped, the parts should be rinsed with clear water. If no soda is available, strong soap suds may be used. If any metal part has been deeply corroded by battery acid, it should be replaced.

Landing Gear and Tail Wheel. The main landing gear, tie strut and tail wheel, and its fastenings should be carefully examined for indicated damage or any defects. Check to see that the tires are properly inflated. Check the operation of the parking brake and foot brakes. The foot brakes must hold the airplane at full throttle. Check to see that the parking brake is not binding. There should be some slack in the parking-brake cable when the parking-brake handle is fully released.

Control Surfaces. The control surfaces should be examined to determine that they operate freely through their full range. Check to see

AIRCRAFT MAINTENANCE AND SERVICE

that the controls are in the NEUTRAL position when the stick and pedals are in that position.

Aircraft Skin. The fuselage, wings, ailerons, elevators, stabilizers, rudders, and all other parts should be examined to see that there is no failure or damage to the covering. All cowling and inspection covers should be securely fastened in place. In cold weather, all frost should be removed from the surfaces of the airplane by brushing or thawing in a warm place. If the temperature is not too low, the frost may be removed by flushing with water. If any part of the airplane covering is wrinkled, the cause of the wrinkling should be determined. Wrinkled covering may indicate severe damage to the airplane structure at or near the wrinkled fabric.

Miscellaneous. All loose articles in the cabin which could possibly foul or jam the controls should be properly secured, stored, or removed. The windshield and other transparent parts of the cabin should be clean and clear. Sliding windows should be examined for proper operation and security. The door and its fastenings should be examined for proper operation and condition. The safety belts should be checked to see that the belting which passes through a buckle at the point of attachment extends at least one inch beyond the buckle. The buckle and emergency release should not be bent. They must operate freely and hold securely. Check the first-aid kit to see that no articles are missing. Check the fire extinguisher to see that it is in its proper place and properly filled.

Instruments. Check all instruments for broken or loose cover glasses, or other visible defects. Tighten loose cover glasses on all instruments, except the air-speed indicator. The air-speed indicator is a pressure instrument and must be replaced if the cover glass is loose. If there are any markings on the cover glass, the glass should be in its correct position before being tightened into place. The cover glasses should be clean. Check to see that the shock-proof mountings are in good condition. Check the instruments for correct pointer indications. Check the compass for discolored liquid and to see that there are no bubbles present.

Engine Controls. Operate all engine controls to see that they are free and operate properly. If there is any slack in the controls or the controls bind in any position, they should be put in proper condition before starting the engine.

THE LIGHT AIRPLANE

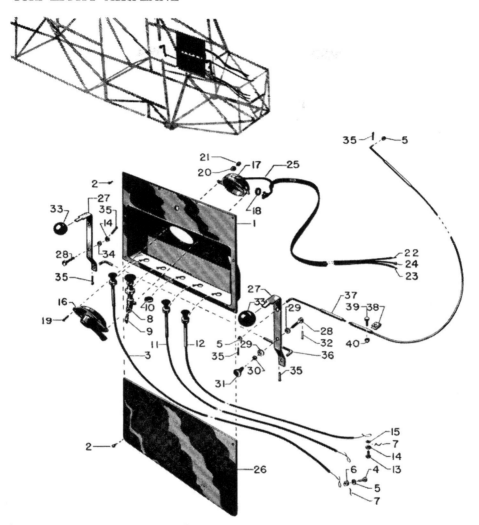

Fig. 66. Power plant controls and control panel showing (1) control panel, (2) stove head screw, (3) cabin heater control, (4) bolt, (5) plain washer, (6) shear nut, (7) safety wire, (8) primer pump, (9) male inverted elbow, (10) plug button, (11) carburetor heater control, (12) mixture control, (13) bolt, (14) shear nut, (15) plain washer, (16) dual ignition switch, (17) ignition switch shield assembly, (18) elastic grommet, (19) roundhead machine screw, (20) lock washer, (21) hex machine screw nut, (22) dual ignition right hand shielded hot wire assembly, (23) dual ignition lefthand shielded hot wire assembly, (24) wire assembly, ignition switch shielded ground, (25) bonding braid, (26) plate and control panel accessory cover, (27) throttle lever, (28) aircraft bolt, (29) throttle lever washer, (30) lock washer, (31) throttle tightening knob, (32) cotter pin, (33) throttle knob, (34) burr washer, (35) cotter pin, (36) throttle connecting rod, (37) throttle control cable, (38) flexible throttle attaching clevis, (39) aircraft bolt, (40) self-locking nut. (Courtesy Taylorcraft Aviation Corporation)

AIRCRAFT MAINTENANCE AND SERVICE

Engine Warm-Up. If any fuel has been spilled on the ground during the refueling or draining of the fuel system, the airplane should be removed to a safe distance before starting the engine. Spilled fuel on the airplane itself should be allowed to evaporate before starting the engine.

The engine should not be started, run-up on the ground, or tested unless a qualified person is seated in the pilot's cockpit. If, in an emergency, the engine must be started without a qualified person in the cockpit, the control stick should be pulled back as far as possible and fastened in that position. The safety belt may usually be used as a fastening. The parking brake should be set, and the wheels chocked. The throttle should be set to within $\frac{1}{4}$ in. of the CLOSED position and locked in that position.

For warm-up or ground testing, the airplane should be headed into the wind. The engine is then started by following the regular starting procedure. If the oil pressure does not start to register within 30 sec. after starting, the engine should be stopped and the trouble located and corrected before the engine is again started.

All engine instruments should be checked for proper operation during the engine warm-up. Each magneto should be checked separately to see that it is operating properly.

If the engine does not stop when the switch is turned off, it should be stopped by turning off the fuel. If this happens, the propeller should not be touched until the trouble has been located and corrected.

After-Flight Inspection. The fuel and oil tanks should be serviced and filled. The propeller should not be touched without first checking to see that the ignition switch is in the OFF position. The propeller should be cleaned and inspected for any damage and for tightness at the hub. The cowling should be removed and a careful check made for indication of leaks in the fuel and oil systems. Wiring, fuel lines, connections, and all attachments should be examined. The exhaust and intake pipes should be checked to see that they are tight. The engine mount should be examined for looseness of fastenings or signs of failure.

Daily Inspection. Each day that the airplane is to be flown, the following inspection should be performed.

Propeller. Be sure that the ignition is in the OFF position and examine the propeller carefully for any damage or looseness.

Engine. Remove upper and lower engine cowling. Inspect the engine mount visually for cracks, dents, bends, and security of fastenings. Bent or cracked tubes are often indicated by checked or loose spots of paint. Check for security of fastening of baffles and any indication of damage. Clean the oil from the inside of the cowling and the front of the fire wall. If there is an excessive amount of oil present, check carefully for oil leaks. Inspect oil drain plugs and oil lines. Check exhaust manifold for blown gaskets, missing bolts or nuts, broken clamps, and general condition. Check fuel lines and carburetor for security of fastenings and for leaking gaskets. Push-rod-housing hose clamps should be snug, but should not be tightened unless signs of leaking are present. Check all engine controls. Examine the air filter and clean it, if necessary. If the airplane is operating under dusty conditions, the air filter should be cleaned after each flight. Normally, the air filter should be cleaned at least once for every 25 hr. of operation. Check for leaks in the fuel system, including the priming system. Check all safetying. Before replacing the cowling, examine it carefully for cracks, distortions, dents, and condition of fasteners.

Landing Gear. Inspect the tires for proper inflation, using a tire gauge. Examine them for cuts or other damage. Check all bolts, nuts, and cotter pins. Check the shock-absorbing devices. Check the brake system to see that the cables are securely fastened and that the brakes operate properly. The tail wheel and its accessories should be carefully inspected. Examine all bolts, nuts, and cotter pins for security. Examine the springs and spring leaves for cracks, breaks, or any signs of failure.

Fuselage. Check the fuselage covering for damage or for wrinkles which may be an indication of structural damage.

Fixed Surfaces. Check the wing surfaces for tears, wrinkles, or other damage. Check all inspection holes, covers, plates, and their fastenings. Check the fairing bands at the point where the wing joins the fuselage for cracks, tightness, and proper position. Check the struts for dents, bends, or other damage. Check the strut fastenings for security and proper safetying. Check the horizontal and vertical stabilizers for damage or wrinkles in the covering. Inspect all safetying on tie rods, hinges, and other attachments. Be sure that the lock nuts on the tie rods are properly tightened.

Control Surfaces. All control surfaces should be checked for holes, tears, or wrinkles in the covering. Inspect the safetying on the ailerons, rudder, and elevator hinge pins. Check the safetying of the turnbuckles

AIRCRAFT MAINTENANCE AND SERVICE

and lock nuts on the control cables and rods. Move the aileron firmly against both the lower and upper stops to test for slippage of the cables that control the aileron. Slippage is indicated if the ailerons stop when coming in contact with the control stops and then go beyond their correct range of travel when additional load is applied. Slippage should be corrected by properly tightening the cable clamps on the large pulleys inside the wings. When tightening the cable pulley clamps, the ailerons and control stick should be in the neutral position. Inspect the cable connections to the rudder and elevator horns.

Cabin. Inspect the cabin for loose objects that might jam the controls. Inspect all parts of the cabin for proper condition. Check the doors and windows for proper operation. Check the safety belts for frays and proper buckle operation. Check the safety belt for proper attachment to the airplane structure. Check the fire extinguisher. Check the proper safetying of all controls located in the cabin. Check the operation of the elevator trim tab control. Check the operation of the parking brake and foot brake. Check the nuts on the brake-cable adjustments which must be tight. Check the frictional adjustments on the engine and other controls. Check the fuel cut-off valves for leaks and ease of operation. Check the primer for leaks and operation. Check the fuel-line connections that are accessible from the cabin. Check all instruments and their attachments and mountings.

VII TECHNICAL DRAWINGS

It is becoming more and more important that the mechanic be able to interpret engineering drawings rapidly and accurately. It is by means of this type of drawing that the workman can duplicate exactly the ideas of the designer. The workman usually receives a copy of the engineering drawings or parts of the original drawings representing the part which he is to build. These drawings are called "working drawings."

Most work is done from blueprints in the form of detailed drawings. There are a number of different ways in which these drawings are furnished to the workman. The blueprint is the most common. On blueprints, the entire background is blue, and lines representing the drawing, dimensions, and figures appear in white.

There is also the brownprint which is just like a blueprint except that the background is brown instead of blue; the lines on this drawing appear in white against a brown background. Blueprints and brownprints are made directly from the tracing, which is a duplication in ink of the original drawing on translucent tracing cloth. The drawings are printed from the tracing cloth in much the same manner that a picture is printed from a negative on sensitive paper to make the finished photograph.

The blue-lined print is a print upon which the lines appear in blue against a white background. There are also brown-lined and black-lined prints.

By means of photography, photostats are obtained. They may be white lines on a black background, which is the negative, or black lines on a white background, which is the positive. This method is used to enlarge or reduce drawings.

Pictorial or Oblique Views. Pictorial or oblique views picture the object in such a manner that three sides of the object are seen. Usually, the sides which appear are the right side or face, the plan or top face,

AIRCRAFT MAINTENANCE AND SERVICE

and the front face or front view. Such a view gives the impression of seeing the whole object. On this type of drawing, the lines are foreshortened or drawn in perspective in order to make the object appear to have three dimensions. In engineering drawings, the three faces

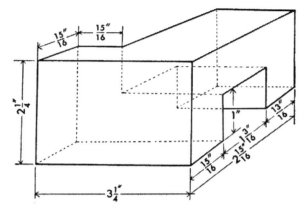

Fig. 67. Pictorial drawing with dimensions.

which appear in the view are drawn as though viewed at right angles to the face. For example, a rectangular block, shown in Figure 68, viewed directly from the front appears as a rectangle. This is called the front view. Viewed from the right side, another rectangle is seen which is drawn to the right of the front view and is called the side view.

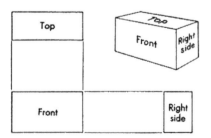

Fig. 68. Three-view drawing of a rectangular block.

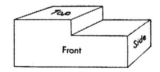

Fig. 69. Pictorial drawing of an object.

Directly above the front view is drawn the third rectangle, which is called the top or plan view. In drawing these three views, it is customary to place the front view in the lower lefthand part of the page with the top view directly above it and the right-side view to the right of the front view. These three views are commonly called a "three-view drawing."

TECHNICAL DRAWINGS

Blueprint Lines. Lines representing the edge of an object which can be seen are drawn heavy. These are called visible boundary lines. To represent an edge which is hidden behind some other part of the object, fine dotted lines are used. These lines are called invisible

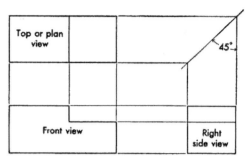

Fig. 70. Three-view drawing of the object shown in Fig. 69.

boundary lines. Sometimes a heavy boundary line on a view may represent both the visible boundary line and the invisible boundary line, where the invisible line is directly behind the visible boundary line.

The beginner usually has some trouble in getting a picture from these three flat views. With practice, however, he can get a mental picture of the complete object from these three views. As he gains

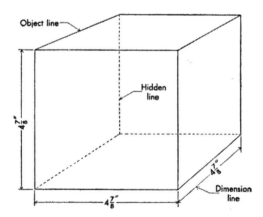

Fig. 71. A pictorial drawing showing object lines and hidden lines.

experience, he can visualize a complicated object from the three views shown in this type of drawing.

The corresponding parts of each view are usually connected by thin solid lines which are called extension lines. Dimension lines show the

AIRCRAFT MAINTENANCE AND SERVICE

distance between various parts of the object illustrated. Between these lines are placed a broken arrow and the dimensions in inches or feet. Dimension lines are illustrated in the drawings used in Figures 67 and 71.

Sections. Sectional views are made by cutting through a part of the object and drawing the area which has been cut. These views are often used to show important details which cannot be otherwise shown on the drawing. Usually a line is made on the front, side, or top view,

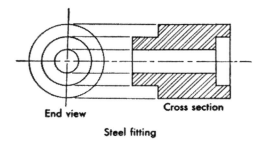

Fig. 72. A typical cross section.

showing where the cut was made, and letters are used to indicate where the cut was made. On the sectional drawing, the same letters are used. A sectional view is usually marked with lines which represent the kind of material cut, which may be wood, steel, brass, etc. These cross lines or markings are called conventional, cross-sectional markings.

A complete drawing or assembly drawing may represent either the complete structure, as for example an airplane; or a part of a complete structure, such as an airplane wing; or show how a smaller fitting is to be made. A detailed drawing shows only one small part in detail. Detailed drawings give all the information needed to complete the part. Each detail is plainly shown on the drawing to enable the workman to complete the part. It would be almost impossible to build an airplane from a complete assembly drawing. Each part is broken down into detailed drawings which usually make it possible for one workman to complete all the work on the part shown on his detailed drawing.

Broken sections are used where it would be impossible to show a very long part on the drawing. For example, a spar may have its root and its tip very close together on the drawing, with the greater portion of the spar left out. This is made clear in Figure 73.

All blueprints and technical drawings follow the same pattern in order that any person familiar with this type of drawing will know

TECHNICAL DRAWINGS

exactly where to look to obtain needed information. In a ruled-off box in the lower righthand corner of the drawing is usually placed the title of the drawing in large letters. This title gives the name of the assembly and the name of the job. On a small detailed drawing, this title may simply give the name of the operation to be performed or the name of a single part to be made. Other information needed to clarify or make plain the title is often placed in this box, or in ruled-off portions of

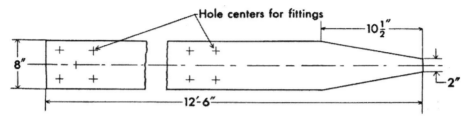

Fig. 73. A broken view showing the root and tip of a spar.

the box. The number of the drawing is placed to the right of the title. In the box will be found the name of the engineer who made the drawing or the company producing the part. The name of the draftsman may appear. If the drawing has been checked by another engineer or draftsman, the name of this person will also be placed in this box. The date of the drawing and the date approved are noted.

All materials used on the job are contained in a box near or above the title box. This box may have the word, "noted," in it which indicates that notes pertaining to the proper material are found elsewhere on the print. On blueprints the various parts are numbered. This enables the workman to get a detailed blueprint or drawing of that part.

Specifications are directions which accompany drawings describing the kinds of material to be used. The driving of an extra nail or the placing of an extra screw might weaken instead of strengthen the part. Every mechanic should have a good understanding of blueprints and drawings. A person learns to read a blueprint much as he learns to read the printed page. A blueprint which looks complicated to the beginner is as readable to a person familiar with the symbols and lines as is the printed matter in a simple school book.

VIII GENERAL SAFETY PRACTICES

The following general safety practices should always be observed by the mechanic.

Consider safety as a part of every job you do.

Take time to do your job the safe way.

Don't engage in any kind of horseplay on company property.

When crossing aisles and passageways, walk with care and keep to the right, especially at intersections. When using stairs, keep hands out of pockets, ready to catch yourself if you should stumble.

Place no tool, material, or other object in any place where it might fall and injure someone or fall into machinery.

Don't toss tools or parts to fellow workmen.

Never walk under a ladder. There is always the possibility of being struck by something from above.

Don't use unsafe ladders or improvised ladders such as packing boxes.

Don't transfer from one ladder to another. Get down.

Danger tags are placed on machinery and other equipment for a definite purpose and are not to be removed.

Handle waste, rags, and industrial towels with care because of possible concealed sharp objects.

Oily waste, rags, and gloves should always be placed in containers provided for this particular purpose. Otherwise they may ignite and cause a fire.

Clean up spilled oil at once, no matter who spilled it.

Unless your job involves repairing and maintaining electrical machinery, wires, and apparatus, do none of this kind of work.

To avoid strains, learn to lift the right way; bend your knees, keep your body erect, then push upward with your legs. It is much easier and safer. Do not try to lift or push objects which may be too heavy for you. Ask for help when you need it.

GENERAL SAFETY PRACTICES

Never wear loose clothing, dangling neckties, or loosely woven clothing, such as sweaters, when working near rapidly revolving machinery. The above is a common cause of accidents.

In the factory, women must wear low-heeled shoes with a closed heel and toe. A durable, lightweight cap with a visor is worn by women working on machines, welding, riveting, or drilling, or doing any other work where a hazard to their hair or scalp exists. Slack suits or coveralls are worn on jobs which require climbing or elevated positions.

Avoid cuts and scratches by wearing suitable gloves.

When piling material, place paper between each sheet.

Don't pile material so that sharp edges protrude.

All parts of power saws are particularly dangerous, and material being sawed should be pushed past the saw with wooden blocks designed for this purpose.

Never let the hands come close to rapidly moving parts such as router heads, band or circular saws, shears, and presses.

Always wear goggles or a face shield when cleaning with air, or when using any type of machine which may throw off particles of metal.

Always wear goggles when working on something overhead.

When using files, care should be taken not to cut the hands on the edge of the material being filed.

Always clean, sharpen, and replace tools immediately after use.

Sharp tools should be covered when not in use.

Report all injuries, no matter how slight.

When driving out a rivet, be sure that there is no one in line with the rivet.

Keep fingers away from the jaws of a rivet squeeze.

Use a nozzle or blow gun on the hose when cleaning with air.

Warn all workmen near you before you start cleaning with air.

Don't disconnect air tools without shutting off the air pressure first.

Sweep up shot spilled from shot bags at once.

Don't take tool boxes on top of wings or center sections.

Don't leave parts on wings or stands where they may fall on men working below.

Replace corner pads and covers when through; think of the other fellow.

Don't start the motor of an extension drill until you are guiding the drill with one hand.

Care should be taken not to mar the work being drilled.

AIRCRAFT MAINTENANCE AND SERVICE

Don't run an extension drill near your hair; you may lose a portion of your hair in an instant.

Don't hold any small object while drilling it; clamp the work in a drill vise.

Don't operate any machine unless authorized to do so.

Don't use dull tools; use sharp ones and select the right tools for the job.

Don't use extension lights without guards; also, check for broken insulations.

Don't use tools with split handles.

Throw away wrenches that slip and don't fit.

Smooth up sharp edges on all jobs where possible.

Don't handle long, sharp-edged parts as if you were the only one present; look around first.

Don't reach into places where mechanism can move, until you take steps to assure safety.

Always report to First Aid with any scratch that draws blood; avoid infection.

Never drill a hole anywhere at any time without looking on the other side first.

Check all measurements carefully before drilling or cutting any material.

Always use a chuck key; it saves time, the drill, and the chuck.

Always see that the screwdriver bit you are using fits the screw you are driving.

Never draw more stock than is required for the job you are doing.

Always report any condition that you think constitutes a safety hazard.

Ask all the questions pertaining to your job that you wish to, regardless of how foolish they may seem.

If you make a mistake, tell your supervisor at once.

Keep your toolbox clear of stock parts and in condition for inspection at all times.

Take time to pick up dropped screws, nuts, and bolts.

Put away light cords and air hoses when through using them and at the end of working hours.

Demand sufficient instruction for doing your job right the first time. Don't guess.

Always replace guards when through working under them.

GENERAL SAFETY PRACTICES

Don't leave an unfinished job without notifying your supervisor.

Screws and bolts must extend one and one-half to three threads through stopnuts.

The standard practice with bolts or screws, whenever possible, is to have the head of the bolt or screw up, outboard, or to the front.

Do not use pliers of any kind on cannon plugs or breeze fittings; tighten by hand.

A screwdriver sharpened like a knife will slip out of the screw slot, hence it will raise a burr or break the screw. Since thousands of screws are used through the assembly of each airplane, the importance of the proper use of properly conditioned screwdrivers cannot be overestimated.

IX GLUES, GLUED JOINTS, AND PLYWOOD

Kinds of Glues. Casein and synthetic resin glues are the most important kinds used in aircraft construction and repair.

Casein Glues. Casein glues, which are prepared largely from the curd of milk, usually contain other material such as lime or sodium salt to help dissolve the casein and make a more desirable mixture of the glue. In using casein glues, the manufacturer's directions for mixing and application should be closely followed.

Mixing Casein Glues. For casein glue, a ratio of one part of dry glue to two parts of water by weight produces a mixture suitable for most joints. In mixing prepared casein glues, the water should first be placed in a mixing bowl and the glue sprinkled or sifted in slowly while stir-

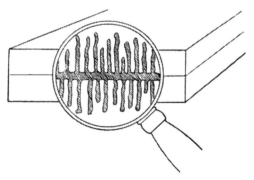

Fig. 74. A magnified glue joint showing glue dowels.

ring rapidly. Figure 75 shows a simple balance which may be made by the workman to obtain the proper mixture of water and glue. There are many types of mechanical mixers which may be used, although for small amounts an ordinary wooden paddle operated by hand may be satisfactory.

When a mechanical mixer is used, the mixing paddle should turn at

GLUES, GLUED JOINTS, AND PLYWOOD

a rate of 100 to 120 revolutions per minute. After all of the glue has been added, the mixer should be slowed down to about one half this rate and stirring continued until all particles of the glue have been dissolved and a smooth mixture results. This mixing usually requires

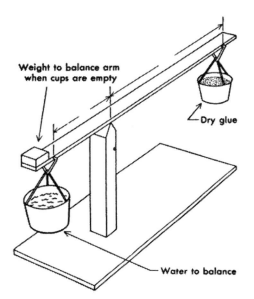

Fig. 75. A home-made balance for weighing water and glue powder.

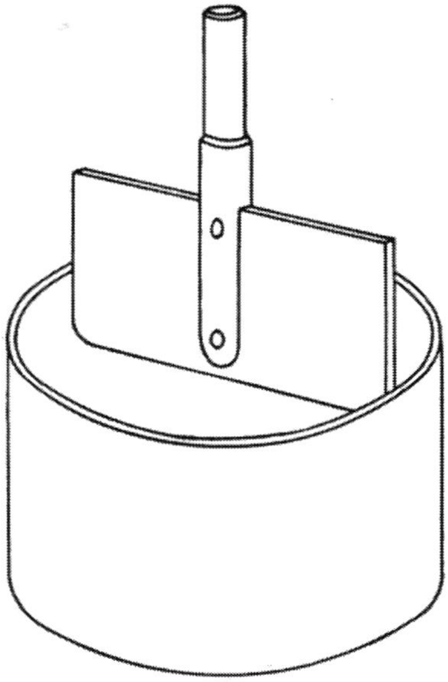

Fig. 76. A home-made drill chuck paddle for mixing glue.

from 20 to 30 min. Most casein glues thicken after this amount of stirring and should be allowed to set for 20 to 30 min. and then restirred. After the second stirring, the mixture should be about the thickness of rich cream. If the mixture shows lumps at the end of the mixing, it should be discarded. The glue should not be stirred rapidly enough to cause foaming. Casein glue should be mixed at not less than 70° F.

A satisfactory mixer may be improvised by placing a small wooden paddle, as shown in Figure 76, in the chuck of a drill press.

Casein glue mixed more than 8 hr. in winter, or 4 hr. in summer should not be used, as it loses its quality.

The proper pressure for glue joints should be strictly followed, and the joint to be glued must be free from any foreign substance which would prevent the glue from entering the pores of the wood.

If the surface cannot be thoroughly cleaned either by planing away damaged wood or foreign material, or by gasoline or carbon tetrachloride, the wood should be discarded. Table I gives the characteristics of

TABLE I. CHARACTERISTICS OF GLUES MOST COMMONLY USED IN CONSTRUCTION AND REPAIR OF AIRCRAFT

PROPERTY OR CHARACTERISTIC	CASEIN GLUE	BLOOD ALBUMIN GLUE	SYNTHETIC RESIN GLUE
Strength (dry)	Very high to high	High to low	Very high to high
Strength (wet, after soaking in water 48 hr.)	About 25% to 50% of dry strength; varies with glue	About 50% to nearly 100% of dry strength	Very high; nearly 100% of dry strength
Durability in 100% relative humidity or prolonged soaking in water	Deteriorates eventually; rate varies with glue	Deteriorates slowly, but completely, in time	Very high if resin is unadulterated
Rate of setting	Rapid	Very fast with heat	Very fast with heat
Working life	Few hours to a day	Few to many hours	Few to several hours for liquid forms; several weeks for films
Consistency of mixed glue	Medium to thick; little change with temperature	Thin to thick; little change with temperature	Medium for liquid forms
Temperature requirements	Unimportant	Heat required to set most glues	Heat required for most glues
Mixing and application	Mixed cold with water; applied cold by hand or mechanical spreaders	Usually mixed cold with water; applied cold by hand or mechanical spreaders	Applied as received or after addition of catalyst; liquid forms best when applied by rubber-covered rolls
Tendency to foam	Slight if not mixed too rapidly	Slight to pronounced	Slight
Tendency to stain wood	Pronounced with certain woods	None, except dark glue may show through thin veneers	None, although glue may penetrate through thin or porous veneers
Dulling effect on tools	Moderate to pronounced	Slight	Moderate
Spreading capacity: Extremes reported Common range	35 to 80 sq. ft. per lb. 40 to 60 sq. ft. per lb.	30 to 100 sq. ft. per lb. ———	30 to 100 sq. ft. per lb. 35 to 50 sq. ft. per lb.

the glues most commonly used in the construction and repair of aircraft.

Gluing Conditions. To produce strong joints, a sufficient amount of glue must be spread on the wood surface. The glue should be of the proper consistency, and the proper pressure should be applied. "Starved" or "crazed" joints may occur where not enough glue is applied. A "dried" joint may occur where the glue is too thick and too much glue has been applied for the amount of pressure. A good joint should have practically no free glue in the joint. A well-glued joint

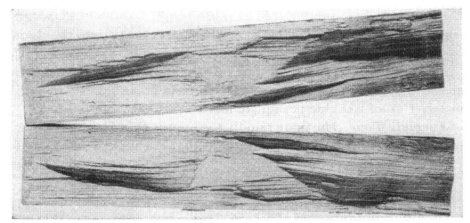
PROPERLY GLUED JOINT

STARVED OR CRAZED JOINT

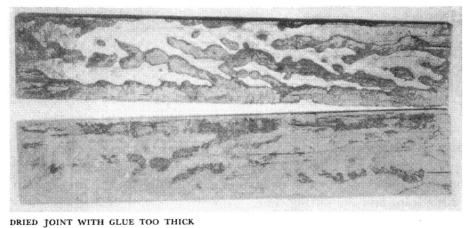
DRIED JOINT WITH GLUE TOO THICK

Fig. 77. Glued joints. (Courtesy Forest Products Laboratory)

consists of a thin layer of glue, most of which is forced into the wood by the pressure applied.

Amount of Glue Required. For casein glue, about 7½ lb. of wet glue will cover approximately 100 sq. ft. of joint surface, or approximately 1 oz. of wet glue per square foot.

Assembly Time. The more rapidly the joint is closed and put under pressure after the glue is applied, the better. For most work, 5 or 10

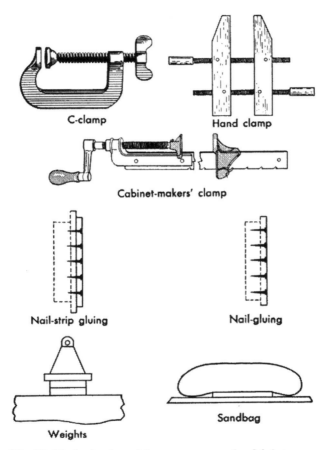

Fig. 78. Methods of applying pressure to glued joints.

min. is all the time which should be allowed to elapse between the spreading of the glue and the application of pressure. A temperature of 70° F., or above, should be maintained throughout the preparation of the joint and the setting time.

Pressure Time. Casein glue should usually be under pressure for not less than 6 or 8 hr.

GLUES, GLUED JOINTS, AND PLYWOOD

Gluing of Different Woods. When the grain of the wood is dense, a smaller amount of glue may be used if the glued surfaces are very smooth and fit closely together.

For softwoods, glue of a medium thickness with a normal spread is recommended. The pressure applied should be from 100 to 150 lb. per sq. in. In gluing, the following woods are usually classified as softwoods: basswood, cottonwood, noble and white fir, western hemlock, ponderosa, sugar and white pine, Port Orford white cedar, western red cedar, redwood, Sitka and white spruce, and yellow poplar.

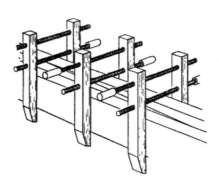

Fig. 79. Solid backing boards prevent slipping of glued joints.

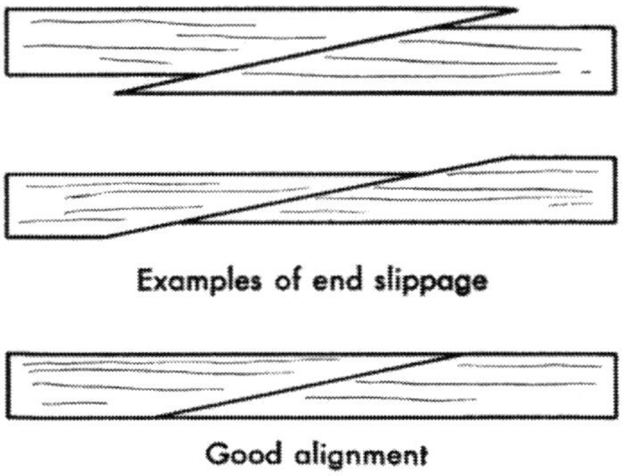

Fig. 80. Examples of scarf joints showing end slippage.

The hardwood group requires a glue somewhat thicker with a normal spread, but the pressure should be from 150 to 250 lb. per sq. in. The following woods are usually classified as hardwoods: white ash, beech, black gum, yellow birch, black cherry, Douglas fir, American elm, rock elm, hickory, mahogany, magnolia, soft maple, hard maple, red oak, white oak, sweet gum, sycamore, black walnut, and water tupelo.

Scarf Joints and End-Grain Gluing. Particular care must be taken on all joints where end-grain wood is glued. All scarf joints are cut at an angle to the direction of the grain and, since they are commonly used in aircraft construction, care must be used whenever gluing this type of joint. Usually joints of this kind should be sized with a glue mixture somewhat thinner than that used for regular gluing. A mixture of 1 part of glue to 3 parts of water has been found to be satisfactory. A sizing coat is a coat of glue which is applied and allowed to dry partially before the joint is glued in the regular manner. The mixture for the final gluing should be somewhat heavier than for ordinary

AIRCRAFT MAINTENANCE AND SERVICE

gluing. About 1 part of glue to 1⅛ parts of water by weight should be used for casein glue. Both surfaces of the joint should be covered with this glue, and approximately 200 lb. pressure per square inch applied. This method should be used wherever end-grain gluing is necessary. Glued joints should be allowed to dry thoroughly so that any moisture added to the wood during the gluing process is evaporated before the final finishing or machining of the joint.

Synthetic Resin Glues. Synthetic resin glues are based on resins made artificially from various chemical components. Phenol-resin (warm- or hot-setting) and urea-resin (cold-setting) glue are the synthetic glues most commonly used in aircraft woodwork. In the preparation of these glues the reaction between the chemicals used in their manufacture is stopped before it is complete. Then when the glue is applied to the parts to be glued, the reaction is completed during the gluing operation. Some glues require an addition of a chemical substance to cause hardening. These substances cause the completion of the reaction. Most plywoods are at present made with the synthetic resin glues by the hot-press method.

Mixing of Synthetic Resin Glues. The manufacturer's directions for mixing these glues should be followed carefully.

Glued Joints. The most important joint used in aircraft construc-

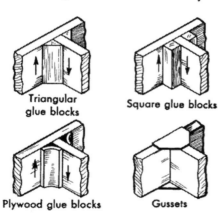

Fig. 81. End-grain joints used in aircraft construction.

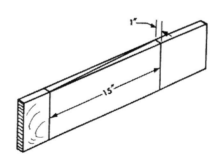

Fig. 82. Marking a board for a scarf joint.

tion and repair is the scarf joint. The scarf joint is made by beveling off at a definite slope the ends or sides of the pieces to be joined together. This slope may be as shallow as 1 in 20, and is rarely steeper than 1 in 10.

GLUES, GLUED JOINTS, AND PLYWOOD

It is sometimes necessary to join the end of a piece of wood to the side of another piece. This is usually done by the aid of glue blocks of the types shown in Figure 81. In wing-rib structures, the joints are strengthened by plywood gussets. Nails are often used in aircraft structures, but are not depended upon to add strength to the structure. They are used to hold the parts together until the glue has set.

Making a Scarf Joint. A scarf joint may be made as follows. Select a board 1″ × 6″ × 30″, and 4 in. from one end draw a line at right angles to one edge. Since the scarf joint is going to have a slope of 1 in 15, draw another line 15 in. from the first line on the same face of the board. Draw lines across the edge of the board, and join the lines on the edge with diagonals, as shown in Figure 82. It is well to extend the lines completely around the board and draw a corresponding diagonal on the opposite edge. The joint should be cut with a fine-toothed saw. Each side of the joint must be finished smoothly, for which a block plane is most satisfactory. Check the feather edge for squareness with a try square. Check the heel for squareness in the same manner. If the feather edge is not square, additional material must be removed. This may make it necessary to remove some material back from the heel because the slope must be 15 in. long when smooth and ready for gluing. In order to support the feather edge, the board may be mounted on

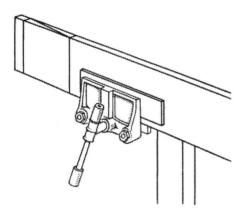

Fig. 83. A board clamped in a vise for sawing a scarf joint.

Fig. 84. Sawing a scarf joint.

another board, as shown in Figure 85. Test the joint surface for flatness by a steel straightedge. The joint should be flat both lengthwise and crosswise. Prepare the other half of the joint in the same manner. If the slope is not exactly true on each part of the joint, the joint must be trued

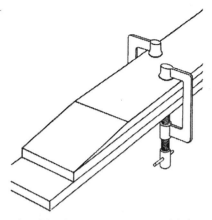

Fig. 85. One half of a scarf joint mounted on a board for smoothing.

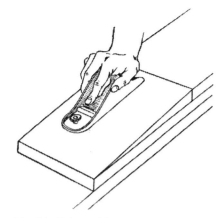

Fig. 86. Using a block plane to smooth a scarf joint.

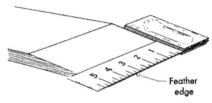

Fig. 87. Testing the feather edge for squareness with a try square.

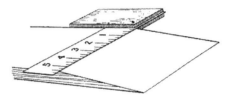

Fig. 88. Testing the heel of a joint for squareness with a try square.

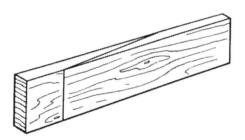

Fig. 89. A finished scarf joint.

or the piece will not be straight after gluing. Figure 77 shows faulty and good joints.

Care must be taken in assembling the joints so that there is no end slippage after the glue has been spread on both surfaces of the joint. The proper glue blocks and clamps are then applied to the joint, as shown in Figure 79. In gluing a joint of this kind, a piece of paper is placed between the material being glued and the glue blocks to prevent the glue's sticking to the blocks when dry.

GLUES, GLUED JOINTS, AND PLYWOOD

Laminations. Laminated woods differ from plywoods in that the grain in the various laminations generally runs in the same direction in each piece. In plywood, the alternate layers have the grain at right angles. The laminations in laminated members are usually much thicker than the veneers used in plywood.

Forming of Plywood. Most plywood for use in the aircraft industry is manufactured and furnished to the aircraft builder in flat sheets. These sheets must be bent or molded to the required form by the aircraft mechanic. The amount that the sheet can be bent is influenced by several conditions, such as thickness, moisture content, grain of the plywood, temperature, quality of the plywood, and the method used in bending. Much plywood can be formed dry without any special treatment. Special treatment consists of soaking, steaming, or placing in hot or boiling water. Where soaking or steaming is applied, care must be taken to be sure that the glues used in forming the plywood will not be affected by the added moisture. Each kind of plywood in

Fig. 90. Rotary cutting of plywood veneers. (Courtesy Curtiss-Wright Corporation)

AIRCRAFT MAINTENANCE AND SERVICE

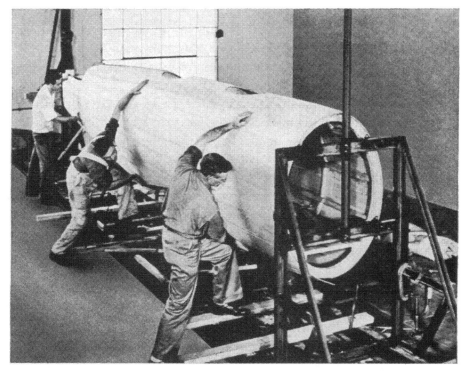

Fig. 91. Assembling a molded plywood fuselage. (Courtesy Timm Aircraft Corporation)

each thickness has a very definite bending radius, beyond which breaking will take place (Table II).

TABLE II. APPROXIMATE MINIMUM RECOMMENDED BEND RADII FOR AIRCRAFT PLYWOOD

PLYWOOD THICKNESS		10 PER CENT MOISTURE CONTENT, BENT ON COLD FORMS		THOROUGHLY SOAKED IN HOT WATER AND BENT ON COLD FORMS	
		Bend at right angles to face grain of plywood	*Bend at 0° or 45° to face grain of plywood*	*Bend at right angles to face grain of plywood*	*Bend at 0° or 45° to face grain of plywood*
Inch	Number of plies	Inches	Inches	Inches	Inches
1/32	3	2.0	1.1	0.5	0.1
1/16	3	5.2	3.2	1.5	0.4
1/8	3	12.0	7.1	3.8	1.2
5/32	3	16.0	10.0	5.3	1.8
3/16	3	20.0	13.0	7.1	2.6
7/32	5	27.0	17.0	10.0	4.0
1/4	5	31.0	20.0	12.0	5.0
5/16	5	43.0	28.0	16.0	7.0
3/8	5	54.0	36.0	21.0	10.0

X DOPES AND FINISHES

Most of the interior finishing of an airplane is done before the parts are assembled. Spar varnish is commonly used for the finishing of all woodwork. Varnish dries by a chemical action on exposure to the air. It should be applied in thin coats to allow proper drying. Any surface which is to be glued should not be covered with varnish or any other finish.

It is necessary to protect metal parts which come in contact with wood from the moisture in the wood. This is done by the application of a suitable sealing material which does not harden completely. Metal parts are covered with special metal sealers or primers to protect them from corrosion.

Dope used for finishing fabric should never be placed over varnished surfaces as it causes the varnish to soften and peel. Dopeproof paint is usually pigmented white, has a resin base, and is thinned with water instead of the usual thinners. For the final finish on exterior surfaces, a pigmented dope or varnish is used. Aluminum powder and lead chromate are two of the common pigments used.

Acetone is usually used as a dope solvent. Ether is sometimes added to dope as a quick-drying thinner.

Wood which is to be covered with doped fabric is usually given a coat of dopeproof material.

Fabric Finishes. There are two general types of finish used on fabrics: a pigmented dope finish, and a semipigmented dope finish.

Pigmented Dope Finish on New Fabric. The following method, if carefully followed, will produce a durable and highly satisfactory pigmented dope finish. After the fabric has been placed on the structure in accordance with approved methods, a coat of clear dope is applied. Be sure that the newly applied fabric is loose enough so that, when it shrinks because of the action of the dope, it will not distort the airplane structure. The dope is applied with a 3- or 4-in. wall brush, and

AIRCRAFT MAINTENANCE AND SERVICE

a full, wet brush of dope should be used. The dope should be applied evenly over the surface. Do not try to brush the dope thin or to cover too much space with one brush full of dope. The first coat will anchor the following coats to the fabric. Remember, this is not painting, but filling the fabric with dope. Saturate the fabric thoroughly with this first coat and allow it to dry from 20 to 45 min., depending on the temperature.

At this point, reinforcing tape, patches, or drain grommets may be applied, either with the second coat or as a separate operation immediately before applying the second coat of clear dope. They are usually applied with the second coat. This is accomplished by doping the area and placing the tape, or grommets, on the wet dope and pressing firmly into place. Be sure no air bubbles are trapped beneath the fabric being applied. The second coat will not require as much dope as the first coat, but a plentiful amount should be used. All coats after the second coat should be sprayed on. If necessary these coats may be brushed on, but should be applied very thinly with a technique similar to applying

Fig. 92. Spraying the finish on fabric-covered wings. (Courtesy Piper Aircraft Corporation)

DOPES AND FINISHES

Fig. 93. Dipping an entire wood wing into the finishing solution. Note the construction of the wing. (Courtesy Forest Products Laboratory)

a thin coat of paint or varnish. Third, fourth, and fifth coats of clear dope are applied. Each coat should dry 30 to 45 min.

Brush coats often require thinning. A nitrate dope thinner of standard brand should be used. Thin only enough to reach the proper brushing quality. Most dope requires some thinning for spraying. However, that used with pressure spraying equipment requires less thinning than for suction equipment. Use as little thinner as necessary.

The fifth coat of dope, after thoroughly drying (several hours) should be sanded lightly, using a No. 000 sandpaper. When sanding dry dope, the structure should be grounded to avoid sparks caused by static electricity. Clean the surface thoroughly and apply two heavy, wet coats of aluminum pigmented dope. The drying time between coats should be 30 to 45 min.

Aluminum pigmented dope should always be used over the clear dope as a protection against destructive rays from the sun. Aluminum pigmented dope makes it possible to obtain a finer finish than with clear dope alone. Approximately 1½ lb. of fine aluminum powder to each 5 gal. of unthinned clear dope should be used. The dry powder should be wet with dope thinner, making a thin paste before adding to the clear dope. Allow 8 hr. drying time after the second coat of aluminum dope is applied. The surface should then be sanded to a smooth finish, using No. 280 wet or dry sandpaper and water. This operation largely determines the final finish. Clean thoroughly, wiping off all dust, dirt, and sanding mud. A sponge and chamois dampened in clean water should be used. Apply three or more coats of pigmented dope by spraying. These coats should be thinned by adding 20 to 30 per cent thinner or reducer. Allow 30 to 45 min. drying time between coats. After drying 8 to 12 hr., the pigmented coats should be sanded lightly with No. 400 wet or dry sandpaper and water, and polished with a rubbing compound.

Removing Dope and Lacquer from Covering. Ordinary types of paint and varnish remover have an acetone base and contain a small quantity of wax to prevent rapid evaporation. Acetone evaporates rapidly, and without wax would not remain on the surface a sufficient length of time to remove the paint. Removers containing wax should not be used on fabric or plywood. Dope will not dry over wax.

The complete removal of dope finish from fabric is best accomplished by placing the surface to be treated as nearly horizontal as possible. Where the surface cannot be arranged in a horizontal position, soft rags may be soaked in the remover and spread over the surface to be treated. This allows the remover to remain in contact a sufficient time to loosen the dope. The dope is wiped and scraped off as the solvent takes effect. When the opposite side of the fabric from which the finish is to be removed is accessible, rags coated with remover are placed on the old finish to be removed and allowed to dry overnight. The opposite side of the fabric is then wet with the remover, thoroughly loosening the finish, and the rags on the outside are pulled off, which will remove practically all of the original finish.

The general method of removing finish is to treat an area approximately a foot square at a time, completely removing the finish before passing on to a new area. Remover should be applied and, as soon as the finish is softened, should be scraped off with a wide, flexible putty knife or some instrument which will not damage the fabric. The fab-

DOPES AND FINISHES

ric is then rubbed with rags soaked in remover to take off the remaining dope.

Where small areas, because of checks and blisters, need to be finished, a clean, soft rag wet with remover may be used. Rub the area with a circular motion, soaking and wiping the dope at the same time. While this area is still soft, build up again with the number of coats originally used. Add 25 to 30 per cent of dope remover to the first coat. This softens the cleaned area and forms a better bond with the new coat.

Dope should never be applied over old surfaces which have been waxed, unless the wax is thoroughly removed with a wax solvent. Oil, grease, and dirt must always be thoroughly removed from the surface before new coats of finish are applied.

Plywood Finishing. When plywood is used as an outside covering, it is sometimes finished by having a coat of dopeproof material placed on the plywood which is then covered with fabric. The fabric is doped and finished with pigmented dope.

When no fabric covering is used, the plywood is thoroughly sealed with a special plywood sealer. A heavy coat is applied which should fill

Fig. 94. A drawing to show construction of an all-metal aircraft. Most metal parts are given a coat of zinc chromate primer before assembly. (Courtesy Bell Aircraft Corporation)

in all minute cracks and crevices in the plywood. This coat should be applied with a brush, keeping the brush thoroughly wet and filled with the sealer. If possible, the structure may be dipped in a tank of sealer which is the most effective method. The least effective method is application by means of a spray. If it is necessary to use a spray, the mixture should be set as wet as possible, and as heavy a coat as practicable should be applied. The second and third coats of sealer should be sprayed on. After thoroughly drying, 2 coats of aluminum pigmented finish should be sprayed on, allowing 30 to 45 min. for drying between each coat. The plywood then may be finished with any regular pigmented plywood finish.

Finishing Steel and Metal Parts. Metal parts require rustproofing on their interiors. Lionoil is usually used for this purpose. Exposed metal parts and fittings should be cleaned of all rust or scale and a zinc chromate primer with about 4 oz. of aluminum bronze per gallon is used. Two or more coats are applied.

Areas of Contact Between Wood and Metal. All interior surfaces of contact between wooden and metal fittings should be marked off and painted by brushing with a generous coat of bituminous paint to which 2 lb. of aluminum paste per gallon have been added.

XI APPROVED REPAIRS

All repairs to certificated aircraft must either be made in a manner approved by the Civil Aeronautics Administration, or be inspected by an inspector of the Administration. The administrator of the Civil Aeronautics Administration has designated who may perform approved repairs: (1) the manufacturer of the aircraft; (2) an approved repair station; or (3) a certificated aircraft mechanic.

Minor Repairs. Aircraft repairs are divided into two classes: minor repairs and major repairs. Minor repairs consist of repairs to nonstructural members such as the cowling, turtle backs, wing and control surface fairings, electrical installations, windshields, and the patching of fabric and the replacing of fabric-covered surfaces involving an area not greater than that required to cover two adjacent ribs. Minor repairs to plywood-stressed covering may be made when the ribs or structural members are not directly affected, but should not exceed 3 in. in any direction. Minor repairs also include the replacement of complete components or units when such parts are supplied by the original manufacturer or are manufactured in accordance with approved drawings.

Major Repairs. Major repairs are complex repairs of vital importance to the airworthiness of an aircraft. They include the replacing, strengthening, reinforcing, and splicing of structural members such as spars or parts of spars, main truss-type beams, the main ribs of wings, compression members, wing or tail surfaces, brace struts, fuselage, longerons, bulkheads, side trusses or horizontal trusses; and repairs to three or more adjacent ribs, or the leading edge of wings covering three or more main ribs, repairs of fabric requiring the covering of more than two adjacent ribs, the repair of damaged plywood-stressed covering exceeding 3 in. in any direction, the repair of portions of skin sheets by making additional seams, or the splicing of skin sheets.

In making any repairs, approved workmanship and techniques must be used. All materials used must be of equal or of better quality than the original material, and the repair must always equal or exceed the original strength.

Emergency Repairs. It must be understood by the workman that emergency repairs should be replaced by permanent repairs as soon as possible.

Emergency Repairs to Skin. Most emergency or temporary repairs are made to the aircraft covering. These repairs are often made to prevent further damage from wind or weather. If the damaged area of plywood is small, fabric may be doped to the plywood to cover the hole. This is not usually recommended if the hole exceeds, in its greatest diameter after trimming, 40 times the thickness of the plywood. This method of repair is not recommended if the hole is in the leading edges or front area of the fuselage where it is subjected to high air pressures. If necessary in this area proper reinforcements must be made.

Holes of small size may be covered with temporary plywood patches. The jagged holes should be trimmed out, and the edge of the patch beveled to reduce resistance to the air.

Damage to leading edges and other parts of the structure subjected to high air pressure may be covered with plywood. The holes should be trimmed out to a smooth line and covered with plywood having the proper strength. Wherever possible, the holes should be trimmed out to such an extent that the edges of the plywood patch rest on supporting members in the inner structure. For example, a hole in the leading edge of a wing should be trimmed out to such an extent that the plywood patch would have a bearing surface on two adjacent ribs. Plywood may be soaked if necessary to aid in bending, using the form ribs or adjacent edges of the leading edge as a form. Plywood should be glued and nailed in place. It is not necessary to match grains, as this repair is only made to enable the airplane to be flown to a base where permanent repairs may be made. Cracked or broken ribs may often be temporarily repaired by nailing on pieces of material which act as splints, or by tacking a piece of plywood over one entire side of a broken rib. In an emergency, a wing bow or other similar structure may be replaced by a curved piece of wood, such as a piece sawed out of solid lumber.

Permanent Repairs. All permanent repairs are made in such a manner that they are expected to become a permanent part of the aircraft and should, therefore, be made at least as strong as the original part and finished in an approved manner. All permanent repairs must be made in accordance with the Civil Air Regulations as established by the Civil Aeronautics Administration.

Welded Repairs. Oxy-acetylene welding is generally considered to be the best suited for repair work on aircraft structural parts. Other kinds of welding must be specifically approved by the Civil Aeronautics Administration.

All parts being repaired by welding should be held in place by suitable jigs to prevent misalignment due to expansion and contraction during the welding process.

The proper size of welding tip should be used, and the filler rods should be of approved material.

A weld should never be filled with solder, brazing metal, additional welding metal, or other material to improve the appearance of the weld.

In repairing a welded joint which has failed, all old welding material should be removed before rewelding. No attempt should be made to reweld an old joint by remelting the weld metal already in place.

Parts which depend upon heat treatment for their strength should not be welded unless they can be suitably heat-treated after welding.

The diameter of the hole drilled to form a rosette weld should be approximately one fourth the diameter of the outer tube through which it is drilled. Rosette welds on certain types of approved joints, where they would ordinarily be used, may be omitted where the inner tube forms a tight fit with the outer tube and is otherwise securely fastened into place.

Members which have become dented at a fuselage station may be repaired by reinforcing with an approved type of finger patch or patch plate. Fuselage members which have become bent, dented, or cracked may be carefully straightened and repaired by using a split-sleeve reinforcement. Dents should not be filled in with weld material or solder. Fuselage members may be replaced or spliced by making the appropriate approved repair.

In making welded repairs on engine mounts, welds of the highest quality must be made because they must resist the continued vibration of the engine. Damaged tubes should, as a rule, be repaired by being reinforced with a tube of larger diameter fitted over the damaged member. These tubes should be welded in place using the fishmouth type of joint and rosette welds. Where the fishmouth joint cannot be conveniently made, a suitable scarf joint may be acceptable. It is particularly important in making engine-mount repairs that the mount is not pulled out of alignment by the welding operation. An engine mount that has received extensive damage should not be repaired and returned to service without being approved by an inspector of the Civil Aeronautics Administration.

Axle assemblies and various parts of landing-gear assemblies have been divided into repairable and nonrepairable types. Before making a welded repair on any part of a landing gear, the mechanic should check to see whether or not the repair is allowable.

Wing and tail surface spars of built-up tubing may be repaired by using any of the splices or methods approved for other tubular structures. Parts of this type which have been heat-treated should usually be replaced instead of being repaired.

In general, parts which can be conveniently replaced should not be repaired.

The parts of an aircraft structure whose strength depends upon cold-working should not be welded. Aircraft parts which have been brazed, soldered, or bronze welded, should not be fusion welded because the brazing and soldering mixtures penetrate the hot metal, weakening the welded joint.

Such parts as aircraft bolts, turnbuckle ends, axles, and other heat-treated alloy steel parts should not be welded.

Aluminum alloys which are in the heat-treated or cold-worked state should not be welded or soldered.

In making any repair on a certificated aircraft, the mechanic must use considerable judgment as to the type of repair to be made, and it is his responsibility to see that every repair is made in an approved manner.

Sheet-Metal Repairs. Extensive repairs to sheet-metal structures should be made by the manufacturer or an approved repair station rated for this type of work. Only a certificated mechanic thoroughly experienced in the work should attempt repairs to metal aircraft structures.

Unconventional fastenings, such as self-tapping screws, driving screws, Lock-scrus, or any other fastening, unless originally used by the manufacturer, may be used only with the approval of the Civil Aeronautics Administration. No aluminum alloy bolts having a diameter of less than $\frac{1}{4}$ in. should be used in the main structure of an airplane. Aluminum alloy bolts may be used only where it is not necessary to remove them in making inspections, or for the purpose of maintenance and

APPROVED REPAIRS

service. Aluminum alloy nuts may not be used in seaplanes, but may be used on steel bolts in shear on land planes providing the bolts are cadmium plated. Self-locking nuts may be used only in main structures where approved by the Civil Aeronautics Administration. No oversized hole should be drilled where there is any danger of decreasing the original strength of the structure. All repairs to such parts as wing-spar cap strips, fins, fuselages, wing longitudinal stringers, or other highly stressed members should be made in accordance with factory recommendations and the approval of the Civil Aeronautics Administration.

When disassembling for repairs, all parts of the structure should be supported to prevent damage. Rivets should be removed by drilling through the head and driving out with a punch, supporting the material to prevent damage. All rivets near a damaged area should be carefully inspected for partial failure. Sample rivets should be removed from nearby areas to determine whether or not the holes have become elongated or the rivets have been partially sheared.

Do not attempt to heat-treat parts after riveting because this will cause warping. Riveted assemblies should not be heated in a salt bath, because small amounts of salt will remain in cracks and crevices and lead to corrosion. A17S–T rivets may be driven "as received," but 17S–T and 24S–T rivets should be heated before driving. Some of the small-sized rivets may be driven as received, but it is a better practice always to reheat before driving. Rivets should be replaced with a size equal to the original rivet. If it is necessary to enlarge the hole to accommodate the next larger size of rivet, it may be necessary to obtain Civil Aeronautics Administration approval.

Rivets should be properly spaced with the proper edge distance. Rivets should never be used in a structure where they will be subjected to direct tension. Direct tension tends to pull the heads off. 17S–T rivets $3/16$ in. in diameter, or less, and 24S–T rivets $1/32$ in. in diameter, or less, may be replaced with A17S–T rivets for general repairs. A17S–T rivets must be $1/32$ in. greater in diameter than the rivets they replace. Hollow rivets or rivets of decidedly different types may not be substituted for solid rivets without the approval of the Civil Aeronautics Administration. Care should be taken in repairs which require riveting that the rivets in the adjacent structure are not loosened by the vibrations set up by riveting. All nearby rivets should be carefully examined after the repair has been completed. Where possible, a rivet squeeze should be used when making repairs. In riveting through tubes or hollow structures, the rivets should be driven only enough to form a small head. To attempt to form a standard head will cause the rivet to buckle within the tube.

All material used in making repairs should, whenever possible, be the same as the original material. Any material substituted must be of the same type of alloy. 24S–T may be substituted for 17S–T, but 17S–T should not be substituted for 24S–T unless a heavier gauge is used. All repairs to 24S–T alloy structures, whenever possible, should be made with the same material as the original.

Care must be taken to determine that the material being used has been properly heat-treated. When doubt exists as to the composition of a material, the proper tests should be made.

Parts which, in the original forming operation have been formed in the annealed condition, may have bends which are impossible to form in the heat-treated condition. Material should not be heated to assist in forming unless reheat treatment restoring its original properties can be carried out.

Bend lines should be made to lie, as nearly as possible, at 90° to the grain of the metal. A large bend radii should be allowed for material in the T-condition. All sheet-metal parts which are not to be painted or otherwise finished should be made from aluminum-coated (Alclad) material. All parts used in repairs should be carefully checked for corrosion, pits, kinks, tool marks, scratches, cracks, or any other defects before being installed.

Damaged aluminum alloy ribs of either the built-up or stamped-sheet type may be repaired by suitable reinforcements. This type of repair is acceptable in small and medium-sized airplanes. Unusual types of repairs require the approval of the Civil Aeronautics Administration.

Small holes or cracks in the metal skin which do not involve damage to the structural members may be patched by covering the hole with a metal plate. These patches must be of such a size to take care of the proper edge distance and rivet spacing. If the damage to the metal skin is extensive or the damaged area overlaps the structural members, such as bulkheads and stiffeners, the entire damaged sheet should be removed. In replacing a sheet, the seams, rivet pattern, and rivet size should be the same as the original. If the original seams are not alike, the stronger of the seams should be copied.

Structural members which have been slightly bent may be straight-

APPROVED REPAIRS

ened cold. However, after straightening, the area should be examined with a magnifying glass for injuries to the material. The straightened parts should then be reinforced. The reinforcement may be less than the original strength, depending upon the amount of damage, as the part retains some of its original strength. If any strain cracks or decided indications of failure are present, the reinforcement should be made equal in strength to the original part. The reinforcement should be made by following the manufacturer's *Administrator of Civil Aeronautics Approved* recommendations. In general, the attachment of all reinforcements should extend beyond the damaged area and be fastened to sound metal.

Local heating to assist in bending or any other forming should not be used on aluminum heat-treated alloys. It is sometimes permissible to use a torch with a large soft flame to anneal slightly non-heat-treated alloys of the work-hardened type. This should never be done if it is possible to anneal in a furnace or a bath. The material should not be heated to a temperature which is above the minimum charring temperature for a resinous pine stick. Splices to stringers and flanges should be made in accordance with the manufacturer's recommendations. These recommendations are usually contained in an *Administrator of Civil Aeronautics Approved* repair manual. Stringers are designed to carry both tension and compression forces. When splicing members of this type, the net cross-section area of the spliced member should, in general, be greater than the area of the section of the part spliced. Standard spacing and edge distance of rivets in splices should be followed. When repairing cracks, small holes of $3/32$ in. or $1/8$ in. should be drilled at the extreme end of the crack to prevent further cracking. All reinforcements should be designed to carry the stresses across the damaged portion and stiffen the joints. When repairing welded or riveted tanks, sealing compounds which are insoluble in gasoline or oil should be used to make a tight joint. All traces of flux should be carefully removed after welding.

Aluminum alloy tubing should not be annealed after forming or at overhaul periods, as is required for copper tubing. Spot-welding structures or assemblies should be repaired at an approved station or by the factory, unless approved repairs may be accomplished by substituting rivets for spot welds. Castings, as a rule, should be replaced and no attempt made to repair such parts. Unless aluminum-coated materials are used, protective coatings should be applied. Where possible, anodiz-

AIRCRAFT MAINTENANCE AND SERVICE

ing and proper priming should be carried out on newly repaired parts. Aluminum alloys in contact with wood or dissimilar materials should be protected from corrosion by one or more coats of a zinc chromate primer. When in contact with wood, the wooden parts should be made thoroughly waterproof.

The general methods of making repairs to sheet-metal airplane structures may be found in the Civil Aeronautics Administration's Manual 18. Any repairs not made in accordance with this manual must be approved by the Civil Aeronautics Administration.

XII WOODS USED IN AIRCRAFT

Aircraft Woods—Their Identification and Uses. The woods used in the different parts of any airplane may vary widely in their properties. Spruce, which is suitable for use in all structural parts, is medium light, elastic, strong, tough, and resistant to vibration. However, it is medium soft, and not suitable for carrying heavy compression or tension loads.

Oak and maple are hard and strong and are used where heavy compression and tension loads must be supported—for example, in reinforcing blocks and propellers.

Filler blocks may be made of basswood which is not a strong wood, but is light, soft, and easily worked.

For fairing, balsa is used because of its extreme lightness.

Table III roughly classifies aircraft woods as to their strength, moisture content, specific gravity, weight, grain slope, annual rings, strength, and durability.

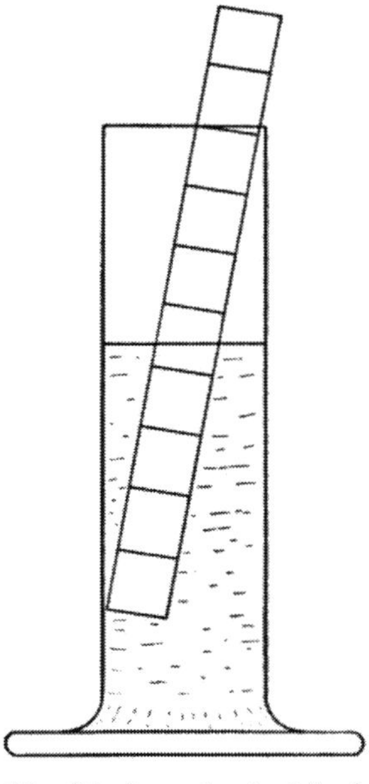

The general properties of the more common aircraft woods and their uses are as follows. The weights given are the average for the species when the moisture content is 15 per cent.

Fig. 95. A method of finding the specific gravity of a wood sample. This figure indicates a specific gravity of approximately 0.45.

Ash. The average specific gravity for white ash is 0.62, minimum permitted 0.56, average hardness 760, and average weight per cubic foot 40 lb. White ash is slightly lighter and softer than oak. It has great strength and stiffness, and bends well. It is particularly suited to construction of bent work, such as wing tips and bulkheads. It is also used in propellers and structural members.

TABLE III. MINIMUM REQUIREMENTS FOR AIRCRAFT WOODS

SPECIES OF WOOD	STRENGTH PROPERTIES AS COMPARED TO SPRUCE	PERMISSIBLE RANGE OF MOISTURE CONTENT PER CENT	SPECIFIC GRAVITY (WHEN OVEN DRY) Average	SPECIFIC GRAVITY (WHEN OVEN DRY) Minimum	POUNDS PER CUBIC FOOT AT 15 PER CENT MOISTURE CONTENT	MAXIMUM PERMISSIBLE GRAIN DEVIATION (SLOPE OF GRAIN)	MINIMUM NUMBER OF ANNUAL RINGS PER INCH	REMARKS
Spruce	100%	8–12%	.40	.36	27	1–15	6	Excellent for all uses. Considered as standard for this table.
Douglas fir	Exceed spruce	8–12	.51	.45	34	1–20	8	May be used as substitute for spruce in same sizes or in slightly reduced sizes providing reductions are substantiated. If used as direct substitute for spruce in same sizes, a reduction in the minimum specific gravity to .38 is permissible. Large solid pieces should be avoided. Gluing satisfactory.
Noble fir	Slightly exceed spruce except 8% deficient in shear	8–12	.40	.36	27	1–20		Satisfactory characteristics with respect to workability, warping and splitting. May be used as direct substitute for spruce in same sizes providing shear does not become critical. Hardness somewhat less than spruce. Gluing satisfactory.
Western hemlock	Slightly exceed spruce	8–12	.44	.40	29	1–20	8	Less uniform in texture than spruce. May be used as a direct substitute for spruce. Upland growth superior to lowland growth. Gluing satisfactory.
White pine	Properties between 85% and 96% of those of spruce	8–12	.38	.34	26	1–20		Excellent working qualities and uniform in properties but somewhat low in hardness and shock-resisting capacity. Cannot be used as substitute for spruce without increase in sizes to compensate for lesser strength. Gluing satisfactory.
White cedar, Port Orford	Exceed spruce	8–12	.44	.40	30	1–20	8	May be used as substitute for spruce in same sizes or in slightly reduced sizes providing reductions are substantiated. Gluing difficult, but satisfactory joints can be obtained if suitable precautions are taken.
Yellow poplar	Slightly less than spruce except in compression and shear	8–12	.45	.38	28	1–20	8	Excellent working qualities. Should not be used as a direct substitute for spruce without carefully accounting for slightly reduced strength properties. Somewhat low in shock-resisting capacity. Gluing satisfactory.

WOODS USED IN AIRCRAFT

Balsa. The average specific gravity is 0.12 to 0.20, and the weight per cubic foot is from 7½ lb. to 12 lb. This wood has but little strength, is extremely light, and is used for fairings and insulation. It sometimes is impregnated with hot paraffin to make it more workable.

Basswood. The average specific gravity is 0.4, minimum permitted 0.36, average hardness 370, and average weight per cubic foot 22.5 lb. Basswood is used in aircraft construction in wing rib nose blocks, veneer for plywood, templates, and filler blocks.

Beech. The average specific gravity is 0.67, minimum permitted 0.6, average hardness 1060, and average weight per cubic foot 42 lb. Beech is used in aircraft construction for plywood veneers and propellers.

Birch. The average specific gravity is 0.59, minimum permitted 0.53, average hardness 1100, and average weight per cubic foot 37 lb. Birch is used in aircraft structures for plywood and propellers.

Cedar (Port Orford). The average specific gravity is 0.44, minimum permitted 0.40, average hardness 520, and average weight per cubic foot 27 lb. Port Orford cedar may be substituted for spruce. Other cedars are not suitable for aircraft construction.

Cherry (Black). The average specific gravity is 0.54, minimum permitted 0.48, average hardness 900, and average weight per cubic foot 33 lb. It is used in aircraft construction for propellers and plywood facing.

Cottonwood. The average specific gravity is 0.43, minimum permitted 0.39, average hardness 410, and average weight per cubic foot 27 lb. Its use in aircraft construction is in veneer for plywood cores.

Elm. The average specific gravity is 0.65, minimum permitted 0.6, average hardness 1230, and average weight per cubic foot 41 lb. This wood is not commonly used in aircraft construction but, when used, is substituted for ash in stringers, longerons, and bent work.

Firs. The average specific gravity is 0.41, minimum permitted 0.37, average hardness 450, and average weight per cubic foot 26 lb. They may be substituted for spruce in aircraft construction but are somewhat heavier.

Fir (Douglas). The average specific gravity is 0.51, minimum permitted 0.45, average hardness 620, and average weight per cubic foot 32 lb. It is used in aircraft construction as a substitute for spruce, being stronger but considerably heavier.

Hemlock. The average specific gravity is 0.44, minimum permitted 0.40, average hardness 650, and average weight per cubic foot 28 lb. It is used in aircraft construction as cores for plywood and for filler blocks.

Hickory. The average specific gravity is 0.8, minimum permitted 0.7, average hardness 1250, and average weight per cubic foot 50 lb. It is useful in aircraft construction for compression blocks and stringers.

Mahogany. The average specific gravity is 0.51, minimum permitted 0.46, average hardness 790, and average weight per cubic foot 32 lb. Mahogany is used largely in aircraft construction for propellers and plywood.

Maple. The average specific gravity is 0.68, minimum permitted 0.6, average hardness 1270, and average weight per cubic foot 42 lb. Maple is used in aircraft construction for propellers, plywood veneer, and engine mounts.

Oak (Red). The average specific gravity is 0.67, minimum permitted 0.62, average hardness 1240, and average weight per cubic foot 42 lb. Red oak is used in aircraft structures for propellers, engine mounts and compression blocks.

Oak (White). The average specific gravity is 0.72, minimum permitted 0.62, average hardness 1240, and average weight per cubic foot 45 lb. White oak is used in aircraft structures for propellers, engine mounts and compression blocks.

Poplar (Yellow). The average specific gravity is 0.43, minimum permitted 0.38, average hardness 420, and average weight per cubic foot 27 lb. Yellow poplar is used in aircraft construction in plywood cores and nose blocks for ribs.

Pine (White). The average specific gravity is 0.37, minimum permitted 0.34, average hardness 380, and average weight per cubic foot 23 lb. White pine is used in aircraft construction for webs, cap strips, corner blocks, fairing strips, patterns, and forms.

Pine (Yellow). The average specific gravity is 0.36, minimum permitted 0.32, average hardness 380, and average weight per cubic foot 26 lb. It may be substituted in aircraft structures for white pine.

Spruce. The average specific gravity is 0.40, minimum permitted 0.36, average hardness 440, and average weight per cubic foot 27 lb. Spruce is used in aircraft construction in many structural parts such as spars, struts, longerons, float and hull structural members, construction of ribs, webs, landing gear, cap strips, stiffeners, flooring, planking, and plywoods.

Walnut (Black). The average specific gravity is 0.56, minimum permitted 0.52, average hardness 990, and average weight per cubic foot

WOODS USED IN AIRCRAFT

35 lb. Its use in aircraft structures is for propellers and outside veneers of plywood.

The Properties and Characteristics of Aircraft Woods. Annual Rings. Tree rings, which are the rings seen on the end of a log when a tree is sawed at right angles to the trunk, are called annual rings. The thickness of the rings, or, as they are sometimes called, the grain, shows how

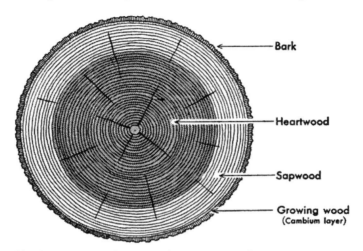

Fig. 96. Heartwood, sapwood, and annual rings.

much wood the tree added during that year. The rings are not the grain. The grain is the direction in which the wood fibers lie.

Heartwood and Sapwood. The older wood is the heartwood or the center of the log. The heartwood is usually darker in color than the wood nearer the bark. There does not seem to be much difference in the strength of wood whether it comes from the heart or near the bark.

The outer layer of the trunk of a tree contains some sap and is called sapwood. The sapwood is usually lighter in color than the heartwood. (See Figure 96.)

Hardwood and Softwood. Wood is divided roughly into two classes: the hardwoods and the softwoods. The hardwoods usually come from the slower growing trees, for example, oak, maple, and ash, while the softwoods come from the more rapidly growing trees such as redwood, fir, and pine.

Physical Characteristics of Wood. The characteristics of wood are divided as follows: durable wood; close-grained and open-grained wood; strong, medium strong, and weak wood; heavy, medium weight and light wood. (See Table IV.)

AIRCRAFT MAINTENANCE AND SERVICE

Fig. 97. Sitka spruce (aircraft wood) on its way to the band saw. (Courtesy American Forest Products Industries)

TABLE IV. CHARACTERISTICS OF SOME COMMON WOODS

KIND	DURABILITY	GRAIN	STRENGTH	HARDNESS	WEIGHT
Birch	Medium	Close	Strong	Hard	Heavy
Hickory	Medium	Open	Strong	Hard	Heavy
Mahogany	Medium	Open	Strong	Hard	Medium
Maple	Poor	Close	Strong	Hard	Heavy
Oak	Good	Open	Strong	Hard	Heavy
Walnut	Good	Open	Strong	Hard	Medium
Cedar	Good	Close	Medium	Medium	Light
Red gum	Poor	Close	Medium	Medium	Medium
Yellow pine	Good	Close	Strong	Medium	Medium
Spruce	Good	Close	Strong	Soft	Light
Basswood	Medium	Close	Weak	Soft	Light
Cypress	Good	Close	Medium	Soft	Light
White pine	Medium	Close	Medium	Soft	Light
Poplar	Good	Close	Medium	Soft	Light
Redwood	Very good	Close	Weak	Soft	Light

WOODS USED IN AIRCRAFT

Quarter Sawing and Flat Sawing. Plain-sawed lumber is obtained when the log is sawed by beginning at one side and sawing off boards. The boards which are sawed from the side of a log have a flat grain which does not wear as well as edge grain. The board which is sawed out of the center of the log has the annual rings almost perpendicular to the face of the board. This board is called "edge grain." Edge grain is sometimes called "vertical grain." (See Figure 98.)

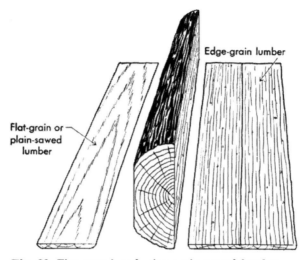

Fig. 98. Flat sawed and edge grain sawed lumber.

Seasoning of Lumber. Most lumber used in aircraft structures is kiln-dried. Kiln drying is accomplished by placing the freshly sawed lumber in large oven-like rooms called kilns.

Unequal shrinkage may cause stresses in the lumber which manifest themselves in such defects as checking, honeycombing, case hardening, or warping. Sometimes the fibers are actually collapsed from too rapid drying. This defect is called "collapse."

Wood that is checked, honeycombed, or collapsed is not suitable for use in aircraft structures.

Moisture Content of Wood. The moisture content of the wood for aircraft structure should usually be between 8 and 12 per cent.

Selection of Aircraft Woods. It is important to know the requirements of wood to be used in all parts of an aircraft structure.

General Requirements. All wood used in aircraft structures should, of course, be sound and free from defects which might cause failure of the part. Wood used in parts which are subjected to bending forces

AIRCRAFT MAINTENANCE AND SERVICE

should be free from decided cross grain, curly grain, spiral grain, sap pockets, checks, splits, and shakes. Any sign of decay should cause rejection of wood for use in aircraft construction.

Wood for Spars and Spar Parts. All wood used in spars or parts of spars, whether the spar is solid or laminated, should be straight grained and sound.

Solid Spars. The material from which a spar is to be made should be straight grained for at least $3/4$ of its depth. Edge-grain lumber should be used for making spars. Small variations in slope which affect less than 5 per cent of the spar may be allowed if the slope is not steeper than 1 in 12. It is sometimes permissible to allow a slope of 1 in 12 near the top of the spar. Any deviation from the standard minimum in parts of the spar near fittings should not be allowed. Knots having a diameter not greater than $1/8$ in. or $1/8$ of the width of the face upon which they appear may be per-

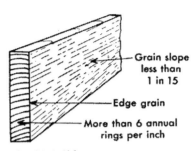

Fig. 99 Solid spar.

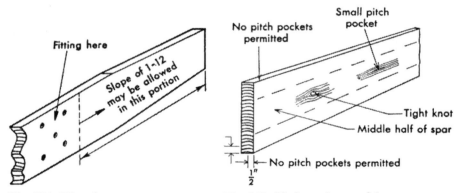

Fig. 100. Tip of spar.   Fig. 101. Pitch pockets and knots.

mitted if firm and sound. Knots should not be permitted at the corners of a spar. Pitch pockets in either edge of the spar should not be permitted. Small pitch pockets in the face of the spar which are less than $1/8$ in. deep and $1/16$ in. wide and less than 1 in. in length may be permitted. Pitch pockets of this type are not permitted if they are closer together than approximately 12 in., neither are they allowed if they occur in the same or adjoining annual rings.

Pitch streaks may be permitted on any face of a spar if the streak is less than $1/2$ in. wide. No pitch streaks should be allowed within 24 in.

WOODS USED IN AIRCRAFT

of the root end of the spar. Spar material must be free from compression wood or compression failure. When necessary, spar material may be spliced, but only one splice should be made in a spar up to 24 ft. in length.

Laminated Spars. When a spar is built up by gluing two or more pieces of wood together in such a manner that the glued surfaces are

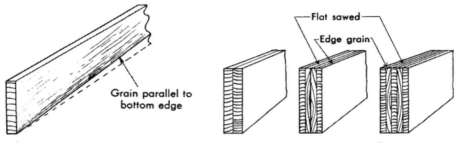

Fig. 102. Grain slope in tapered spar.

Fig. 103. Vertically laminated spars.

parallel to the face of the spar, the spar is called a vertically laminated spar. The glued surfaces run from top to bottom of the spar, as shown in Figure 103. In building laminated spars, the grain should be arranged in the laminations so that it runs in different directions in the adjoining pieces. Wood selected for vertically laminated spars should meet the same requirements as for solid spars.

Solid Spar Flanges. A spar flange is formed by gluing a strengthening piece of material along the upper or lower edge of a spar. The requirements for material for spar flanges are the same as those for solid spars.

Laminated Spar Flanges. If a spar flange is built up by gluing two or more pieces of wood together, it is called a laminated spar flange. If laminated flanges exceed 2½ in. in width, the flanges should be edge grain on their horizontal face. (See Figure 105.) For example, if

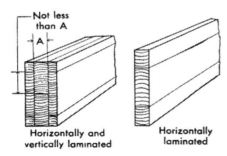

Fig. 104. Horizontally and vertically laminated spars.

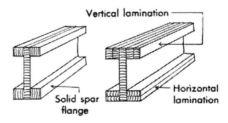

Fig. 105. Solid and laminated spar flanges.

AIRCRAFT MAINTENANCE AND SERVICE

they are horizontally laminated, edge-grain lumber should be used. The requirements for wood to be used in vertically laminated flanges are the same as for solid spar material.

Stressed Parts Having Small Cross Sections. Such parts as cap strips and cross members of ribs have small cross sections compared to their length. The highest grade of lumber available should be used in forming these parts.

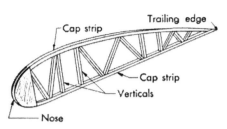

Fig. 106. Wood rib showing stress having small cross sections.

Fuselage Structural Members. Wood used in such fuselage members as bulkheads, stringers, longerons, compression members, and reinforcing blocks should be of high-grade material.

Miscellaneous Wood Parts. Lower-grade materials than those previously described may be used for filler blocks, stiffeners, glue blocks, and doorframes.

It is necessary to use considerable judgment in determining whether or not materials for these purposes are suitable. If the material is to be bent and glued to form curved, laminated parts, it should be free

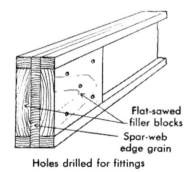

Fig. 107. Filler blocks.

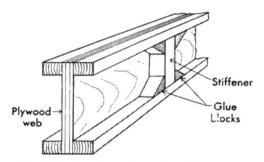

Fig. 108. A plywood web, stiffener, spar flanges and glue blocks.

from any defects such as cross grain, spiral grain, knots, pitch pockets, pitch stains, or any sign of decay.

Defects in Aircraft Woods. Every woodworker, whether he is building a new structure or making repairs, must be an inspector of wood. He must be able to detect such imperfections as grain irregularities, knots, checks, decay, and compression failure.

Grain. There are several grain defects, or variations in grain, which

WOODS USED IN AIRCRAFT

TABLE V. LIMITS OF RINGS PER INCH FOR SPAR STOCK

SPECIES	MINIMUM RINGS PER INCH
Cedar, Port Orford	8
Fir, Douglas	8
Fir, noble	6
Hemlock, western	6
Pine, eastern white	6
Pine, sugar	7
Pine, western white	6
Poplar, yellow	8
Spruce, red	6
Spruce, white	6
Spruce, Sitka	6

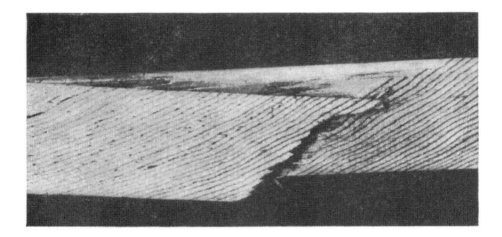

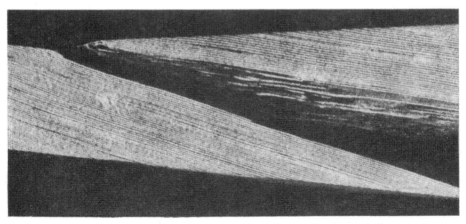

Fig. 109. A typical cross-grain break. (Courtesy Forest Products Laboratory)

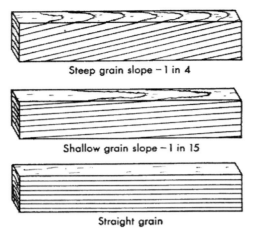

Fig. 110. Grain slope.

Fig. 111. Diagonal grain in plywood shown by annual rings on the surface of the wood. (Courtesy Forest Products Laboratory)

WOODS USED IN AIRCRAFT

may cause an apparently sound piece of wood to be rejected for aircraft use. The direction of the grain is very important.

Slope of grain is given as the number of inches it varies from a center line in a given distance. For example, if the grain deviates 1 in. in 10 in., the slope is 1 in 10. Steeper slopes would be less than 1 in 10 —for example, 1 in 5. A slope of 1 in 20 is not as steep as a slope of 1 in 10.

Grain may take a spiral or twisting course. This is called spiral grain.

Fig. 112. Spiral grain shown by the appearance of the annual rings on the surface of the board. (Courtesy Forest Products Laboratory)

Fig. 113. Ink flow shows grain direction.

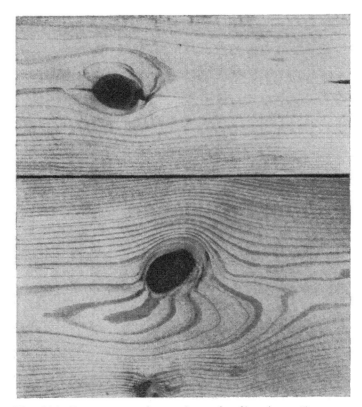

Fig. 114. Knots cause change in grain direction. (Courtesy Forest Products Laboratory)

WOODS USED IN AIRCRAFT

Diagonal grain can usually be determined by the appearance of annual rings along the smooth surface of the board as shown in Figure 112.

Some grain may be curly or twisting. Grain direction can usually be determined by dropping a little free-flowing ink on the surface of the wood. The ink will flow along the grain. (See Figure 113.) In pieces which will be subjected to great stresses, a slope greater than 1 in 20 should not be used.

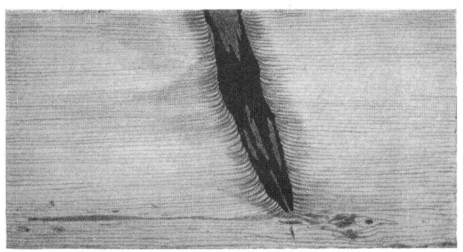

115a. Spike knot.

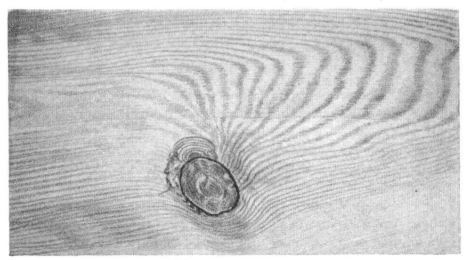

115b. Enclosed knot.

Fig. 115a and b. Enclosed knot and a spike knot. Knots cause a change in the grain direction.

AIRCRAFT MAINTENANCE AND SERVICE

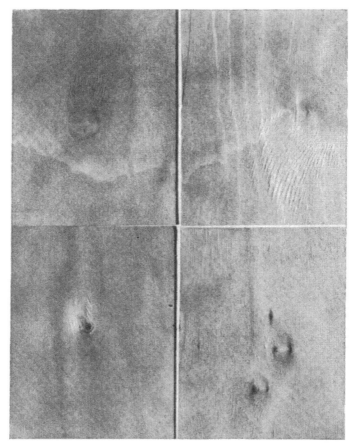

Fig. 116. Burls and knots. (Courtesy Forest Products Laboratory)

Knots. The chief objection to knots is the local effect they have on grain. Knots cause the grain to vary from a straight line, as shown in Figure 114. Table VI shows allowable defects and defects which are not allowable, to assist in determining whether or not the piece of wood containing the knot may be safely used.

Compression Wood. Compression wood is heavy, weak, and subject to extreme shrinking. Figure 117 shows compression wood as it appears in the end of a log and in the board after sawing. Compression wood is not allowed in aircraft structural parts which must carry heavy loads.

Compression Failures. Compression failure is buckling of the wood fibers. This type of failure is difficult to detect without the use of a microscope. This defect usually appears as irregular, threadlike lines

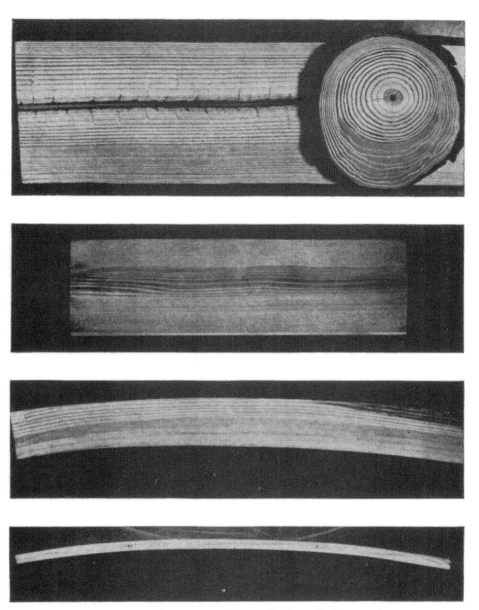

Fig. 117. Compression wood. (Courtesy Forest Products Laboratory)

AIRCRAFT MAINTENANCE AND SERVICE

across the grain. Under the microscope the fibers of wood appear to have been crushed endways. Figures 118 and 119 show compression failures.

Checks, Shakes, and Splits. Checks, shakes, and splits sometimes resemble each other rather closely. They all seriously lower the strength of the wood, and the defect is pronounced.

Fig. 118. Compression failure. (Courtesy Forest Products Laboratory)

A check is a crack crossing the annual rings.

A shake is a crack appearing between the annual rings.

A split is a lengthwise separation of the wood resulting in the tearing apart of the wood cells.

Pitch Pockets. Pitch pockets are openings in the wood, usually occurring between the annual rings, which may resemble shakes. These open-

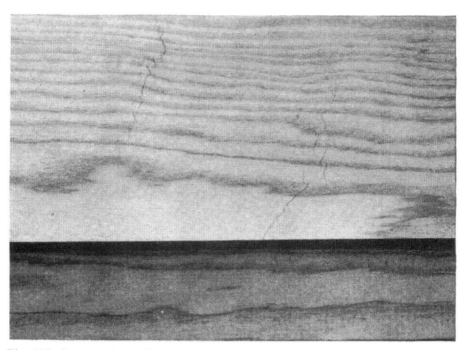

Fig. 119. Compression failure. (Courtesy Forest Products Laboratory)

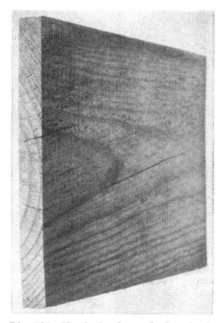

Fig. 120. Checks in the end of a plank. (Courtesy Forest Products Laboratory)

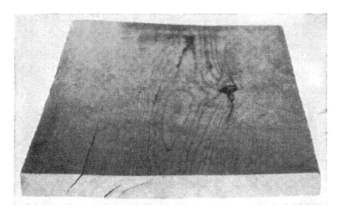

Fig. 121. Shakes in the end of a plank. (Courtesy Forest Products Laboratory)

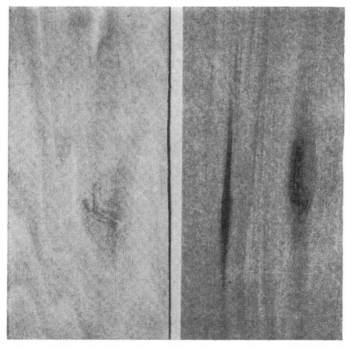

Fig. 122. (Left) Injury to wood during growth. (Right) Bark pockets. (Courtesy Forest Products Laboratory)

WOODS USED IN AIRCRAFT

ings are filled with resin or pitch which may be either solid or semi-liquid.

Bark Pockets. Bark pockets always weaken the wood and are objectionable.

Pitch Streaks. Pitch streaks appear as thick annual rings and should not be confused with pitch pockets. These streaks are objectionable

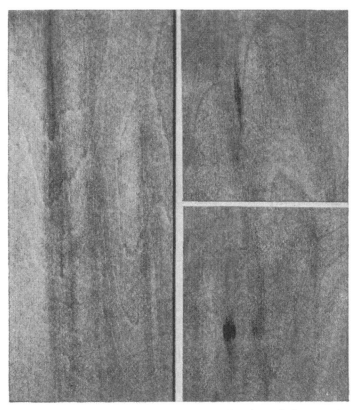

Fig. 123. Pitch streaks. (Courtesy Forest Products Laboratory)

because they add weight to the wood but not strength, and interfere with proper gluing and finishing.

Stains. Any wood which is discolored should be examined very closely and not used if it is possible to obtain stain-free wood. Mineral streaks are defects caused by injury to the wood during its growth. Bird pecks, burls, and lightning cause defects which may or may not be severe enough to cause rejection of the material.

Worm or Insect Holes. When a piece of lumber contains holes caused by worms or insects it should be rejected.

AIRCRAFT MAINTENANCE AND SERVICE

Honeycombing. Figure 125 shows the end of an oak plank which has been honeycombed by too rapid drying.

Collapse. Any indication of collapse, as shown by grooves in the surface of lumber, is cause for rejection. Figure 126 shows the effect of collapse on samples of wood.

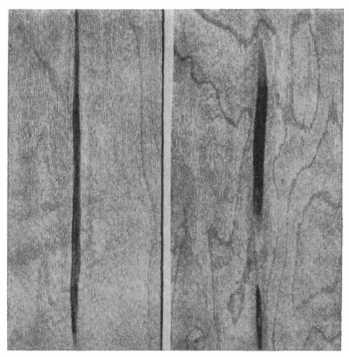

Fig. 124. Mineral streaks in maple. (Courtesy Forest Products Laboratory)

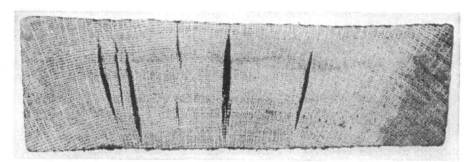

Fig. 125. End of oak plank showing honeycombing. (Courtesy Forest Products Laboratory)

Fig. 126. Collapse in various samples of wood. (Courtesy Forest Products Laboratory)

AIRCRAFT MAINTENANCE AND SERVICE

Brashness. Brashness is a defect in lumber which is caused by severe temperature changes. This wood is brittle due to sudden shrinking of the fibers. Brashness is cause for rejection.

Case Hardening. When lumber is kiln dried too rapidly, unequal shrinking causes internal stresses, and case hardening results. Figure 128 shows tests for case hardening. Case hardening may or may not be a cause for rejection, depending upon the degree permissible for the intended use.

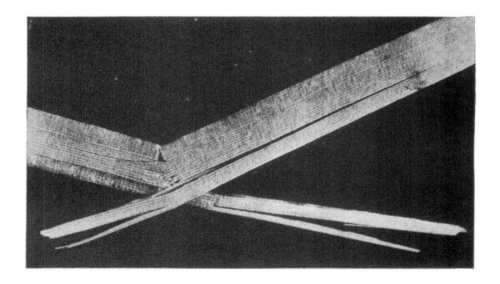

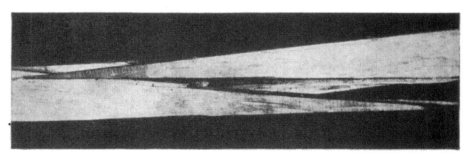

Fig. 127. A typical break of brash wood. (Courtesy Forest Products Laboratory)

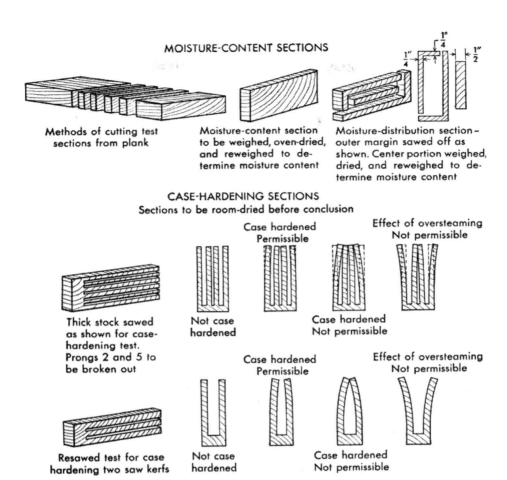

Fig. 128. Case hardening tests. (Courtesy Forest Products Laboratory)

TABLE VI. DEFECTS PERMITTED AND NOT PERMITTED

1. DEFECTS PERMITTED

   (a) *Cross Grain.* Spiral grain, diagonal grain, or a combination of the two is acceptable providing the grain does not diverge from the longitudinal axis of the material more than specified. A check of all four faces of the board is necessary to determine the amount of divergence. The direction of free-flowing ink will frequently assist in determining grain direction.

   (b) *Wavy, Curly, and Interlocked Grain.* Acceptable if local irregularities do not exceed limitations specified for spiral and diagonal grain.

   (c) *Hard Knots.* Sound hard knots up to $3/8$ in. in maximum diameter acceptable providing (a) they are not in projecting portions of I-beams, along the edges of rectangular or beveled unrouted beams, or along the edges of flanges of box beams (except in lowly stressed portions); (b) they do not cause grain divergence at the edges of the board or in the flanges of a beam more than specified; and (c) they are in the center third of the beam and are not closer than 20 in. to another knot or other defect (pertains to $3/8$-in. knots — smaller knots may be proportionately closer). Knots greater than $1/4$ in. should be used with caution.

   (d) *Pin-Knot Clusters.* Small clusters acceptable providing they produce only a small effect on grain direction.

   (e) *Pitch Pockets.* Acceptable in center portion of a beam providing they are at least 14 in. apart when they lie in the same growth ring and do not exceed $1\frac{1}{2}$ in. length $\times$ $1/8$ in. width $\times$ $1/8$ in. depth and providing they are not along the projecting portions of I-beams, along the edges of rectangular or beveled unrouted beams, or along the edges of the flanges of box beams.

   (f) *Mineral Streaks.* Acceptable providing careful inspection fails to reveal any decay.

2. DEFECTS NOT PERMITTED

   (a) *Cross Grain.* Not acceptable unless within limitations noted in 1a.

   (b) *Wavy, Curly, and Interlocked Grain.* Not acceptable unless within limitations noted in 1b.

   (c) *Hard Knots.* Not acceptable unless within limitations noted in 1c.

   (d) *Pin-Knot Clusters.* Not acceptable if they produce large effect on grain direction.

   (e) *Spike Knots.* These are knots running completely through the depth of a beam perpendicular to the annual rings and appear most frequently in quarter-sawed lumber. Wood containing this defect should be rejected.

   (f) *Pitch Pockets.* Not acceptable unless within limitations noted in 1e.

   (g) *Mineral Streaks.* Not acceptable if accompanied by decay. (See 1f.)

   (h) *Checks, Shakes, and Splits.* Checks are longitudinal cracks extending, in general, across the annual rings. Shakes are longitudinal cracks usually between two annual rings. Splits are longitudinal cracks induced by artificially induced stress. Wood containing these defects should be rejected.

   (i) *Compression Wood.* This defect is very detrimental to strength and is difficult to recognize readily. It is characterized by high specific gravity, has the appearance of an excessive growth of summer wood, and in most species shows but little contrast in color between spring wood and summer wood. In doubtful cases, the material should be rejected or samples should be subjected to a toughness machine test to establish the quality of the wood. All material containing compression wood should be rejected.

   (j) *Compression Failures.* This defect is caused from the wood being overstressed in compression due to natural forces during the growth of the tree, felling trees on rough

WOODS USED IN AIRCRAFT

TABLE VI. DEFECTS PERMITTED AND NOT PERMITTED (*continued*)

or irregular ground, or rough handling of logs or lumber. Compression failures are characterized by a buckling of the fibers that appears as streaks on the surface of the piece substantially at right angles to the grain, and they vary from pronounced failures to very fine hairlines that require close inspection to detect. Wood containing obvious failures should be rejected. In doubtful cases, the wood should be rejected or further inspections in the form of microscopic examination or toughness tests made, the latter means being the more reliable.

(k) *Decay.* All stains and discolorations should be examined carefully to determine whether or not they are harmless stains or preliminary or advanced decay. All pieces should be free from rot, dote, red heart, purple heart, and all other forms of decay.

XIII METALS USED IN AIRCRAFT

The mechanic should always be able to determine the kind of metal which he is to weld. He must also know whether or not the metal has been heat-treated, cold-worked, or annealed. When possible, this information should be obtained from blueprints and specifications furnished by the manufacturer.

The identification of metals may be made by a number of comparatively simple tests. Exact identification can be made only by extensive chemical tests and microscopic examinations.

The following will assist the mechanic in identifying most of the metals which he will be called upon to weld:

1. The general appearance of the metal.
2. The appearance of the fractured surface of the metal.
3. The appearance of the freshly polished surface.
4. The chipping of the metal with a cold chisel.
5. The spark test.
6. The action of the metal while being melted under the welding flame.
7. The weight of the metal.
8. The hardness of the metal.

The fractured surface should be closely examined to determine its crystalline structure, roughness, and general appearance.

The newly machined or filed surface should be examined for color, smoothness, and luster.

In making the chip test, a small amount of material is removed from the edge of the sample with a sharp cold chisel. The material removed will vary from small broken fragments to a continuous strip. The chip may have smooth sharp edges, be coarse grained, fine grained, or have sawlike edges where cut. The size of the chip is of importance in iden-

METALS USED IN AIRCRAFT

Fig. 129. Fabricating a metal wing rib by the use of a rivet machine.

tifying the sample. The ease with which chipping takes place also assists in the identification.

The spark test is made by holding a sample of the material against a power grinder. The sparks given off, or the lack of sparks, assist in identifying the sample. The length of the spark stream, its color, and the type of sparks are items which should be noted. Figure 130 shows

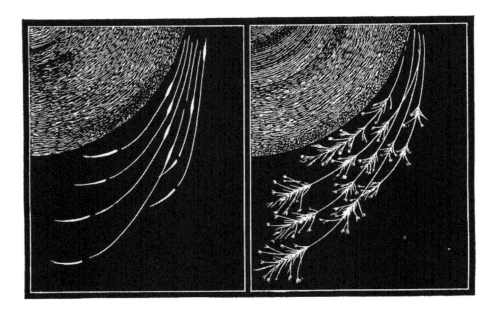
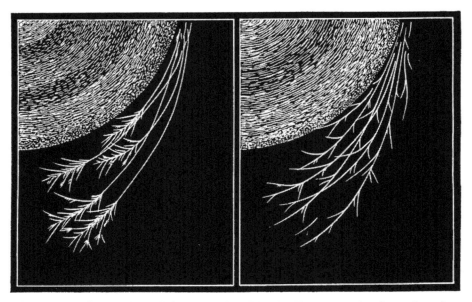

Fig. 130. Fundamental spark forms produced by holding a sample of metal against a power grinder. Upper left: shafts, buds and appendages; upper right: sprigs and sparklers; lower left: shafts and forks; lower right: forks.

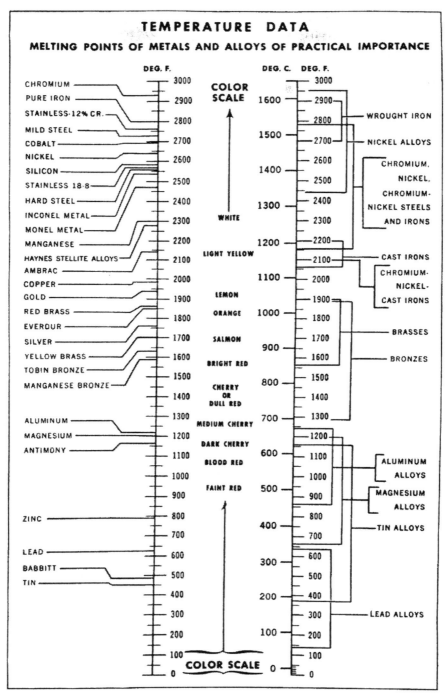

Fig. 131. Temperature data. (Courtesy Linde Air Products Company)

the various types of sparks obtained by the use of a grinding wheel.

The rate of melting from the cold state should be determined by taking a small amount of the material and melting it with the blowpipe. It is sometimes possible to melt a small area on the part to be welded without damage to the parts.

Color changes while heating, the action of the slag formed, the appearance of the molten puddle, and the action of the molten puddle under the flame will all aid in the identification of the metal.

The difference in hardness between metals and alloys can be estimated by the use of a sharp, flat file. If the material is cut by the file with extreme ease and the metal tends to clog the spaces between the file teeth, it is classified as very soft and not heat-treated. If the material offers some resistance to the cutting action of the file and tends to clog the file teeth, it may be classified as soft. If the material offers considerable resistance to the file, but can be removed by a considerable effort, the material may be classified as hard and it may or may not have been heat-treated. If the material can be removed by extreme effort and in small quantities by the file teeth, it may be classified as very hard and has probably been heat-treated. If the file slides over the material and the file teeth are dulled, the material may be classified as extremely hard and it has been heat-treated.

White Cast Iron. On a newly fractured surface, white cast iron shows a fine, silky, silvery white, crystalline appearance. The outside of the

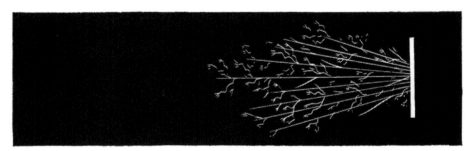

Fig. 132. Sparks from white cast iron produced by a grinding wheel. Note the many sparklers and sprigs.

material shows evidence of a sand mold and is a dull gray. When freshly filed, it shows a silvery white surface. Chips come off in small broken fragments. It is brittle, and the chipped surface is not smooth. The length of the spark stream is approximately 20 in. The volume of sparks is small with many sparklers. The sparklers are small and re-

METALS USED IN AIRCRAFT

peating. The color of the spark stream close to the wheel is red, while the outer end of the stream is straw colored.

Gray Cast Iron. The newly fractured surface of gray cast iron is somewhat the color of white cast iron, but a little darker. The outside shows evidence of a sand mold and is dull gray. A freshly filed surface

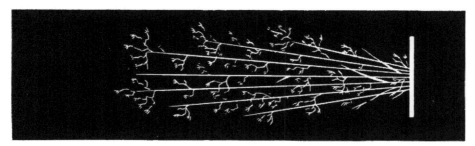

Fig. 133. Sparks produced from gray cast iron by a grinding wheel.

is fairly smooth and light silvery gray. The chips are about $\frac{1}{8}$ in. in length. It is not easily chipped as the chips break off, preventing a smooth cut. Spark tests usually develop a spark stream about 25 in. in length, small in volume, with fewer sparklers than white cast iron. The sparklers are small and repeating, and part of the stream is straw colored.

Malleable Iron. Malleable iron shows a rather fine crystalline, dark gray surface when freshly broken. There is evidence of a sand mold, and the outside is dull gray. The freshly filed surface is smooth and light

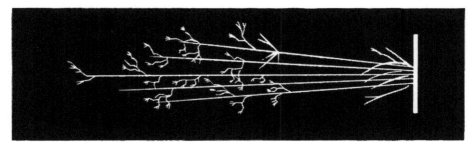

Fig. 134. Sparks produced from malleable iron by a grinding wheel. Note many sprigs of sparklers toward end of stream.

silvery gray. The chips are not as small as those of cast iron, being from $\frac{1}{4}$ to $\frac{3}{8}$ in. in length. Malleable iron is tough and hard to chip. The spark test shows a stream of sparks about 30 in. in length, of moderate volume, with many sparklers toward the outer end of the stream. The sparklers are small and repeating. The entire stream is straw colored.

AIRCRAFT MAINTENANCE AND SERVICE

Wrought Iron. The surface of wrought iron when freshly broken is a bright gray and quite smooth. The unfinished surface is smooth and light gray. The freshly filed surface is smooth and light silvery gray. The chips have a smooth edge where cut and may be continuous if

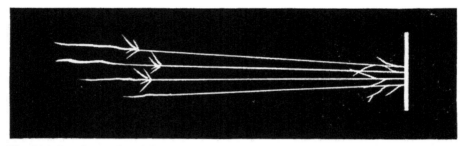

Fig. 135. Sparks produced from wrought iron by a grinding wheel. Note forks and sprigs or sparklers toward end of stream.

desired. Wrought iron is soft and easily cut or chipped. The spark stream is about 65 in. in length, large in volume, with few sparklers. The sparklers appear toward the end of the stream and are forked. The stream next to the grinding wheel is straw colored, while the outer end of the stream is white.

Low-Carbon and Cast Steel. On the freshly broken surface, low-carbon steel is bright gray with a rather fine crystalline appearance. The unfinished surface is dark gray and may have forging marks or

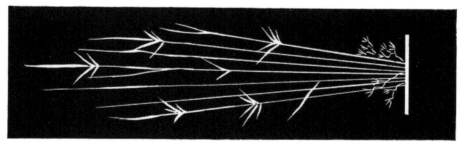

Fig. 136. Sparks produced from low carbon and cast steel by a grinding wheel. Note long shafts with forks.

evidence of a mold. The newly filed surface is smooth and a bright silvery gray. Chips have smooth edges where cut and may be continuous if desired. It is easily cut or chipped. The spark stream is about 70 in. in length; the volume is moderately large. The few sparklers which may occur any place in the spark stream are forked. The color of the spark stream is white.

METALS USED IN AIRCRAFT

High-Carbon Steel. On the newly broken surface, high-carbon steel is light gray and has a fine crystalline appearance. On the unfinished surface, rolling or forging lines may be noticeable. The newly filed surface is smooth and bright silvery gray. The chips show fine grain

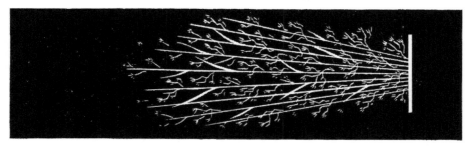

Fig. 137. Sparks produced from high carbon steel by a grinding wheel. The stream is large in volume with many sprigs or sparklers.

fracture, and the edges are lighter in color than low-carbon steel, and may be continuous if desired. The metal is hard and can be chipped. The spark stream is white, about 55 in. in length, and large in volume with many sparklers. The sparklers are small and repeating, and some of the shafts may be forked.

Alloyed Steels. Alloyed steels vary so much in composition that the results of the various tests are not definite. The welding operator should experiment with various known alloys until he is familiar with the appearance of the various tests. For example, Inconel and 18–8 stainless steel are alloys containing nickel and iron which might be easily confused. To determine whether a sample of metal is Inconel or 18–8 stainless steel, a small clean spot on the metal is treated with a few drops of a solution consisting of 10 g. of cupric chloride in 100 cu. cm. of hydrochloric acid. One drop of the solution on the cleaned area should be allowed to remain in contact with the metal for 2 minutes. After 2 minutes, add with a medicine dropper, one drop at a time, 3 or 4 drops of water. If the sample is 18–8 stainless steel, the copper in the cupric chloride solution will be deposited on the metal, leaving a copper-colored spot. If the sample is Inconel, a white spot will be left where the solution contacts the metal. Copper will not be deposited until water is added to the dilute mixture.

Stainless Steel. Stainless steel on a newly broken surface is medium gray, and the grain is rather fine. The unfinished surface is usually dark gray and slightly rough. Rolling or forging lines may be noticeable. The newly filed surface is smooth and a bright silvery gray. The chips vary with the alloys. The spark test for stainless steel produces a stream approximately 50 in. long, of moderate volume, with a few sparklers. The sparklers are forked. The stream next to the grinding

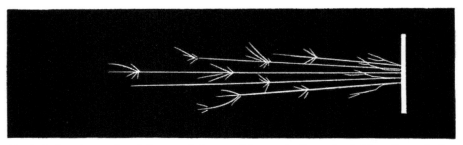

Fig. 138. Sparks produced from stainless steel by a grinding wheel. Note long shafts, forks, few sparklers or sprigs.

wheel is straw colored, while the outer end is white. The action under the flame varies with the composition of the alloys.

Copper. Copper is a bright red color on the newly broken surface, while the unfinished surface may be reddish brown to green, due to corrosion. The surface is usually smooth. The freshly filed surface is a bright copper color which dulls with time. The chips are smooth, have saw edges where cut, and can be continuous if desired. Copper cuts easily and has no spark on the grinding wheel.

Brass and Bronze. The color of brass and bronze may vary from red to yellow. The newly broken surface is quite rough and crystalline. The unfinished surface shows various shades of reddish yellow, yellow green, or brown, and is usually smooth. The freshly filed surface varies from reddish yellow to yellow white and is smooth and bright. The chips are smooth and have saw edges where cut. It is difficult to obtain a continuous chip. There is no spark when the grinding wheel is used.

Aluminum and Aluminum Alloys. Aluminum and aluminum alloys show a fine, crystalline, white surface when freshly broken. The white or light color and extremely light weight make aluminum easy to identify. These alloys, however, should not be confused with alloys of magnesium. Aluminum alloys are usually harder and slightly darker in color than pure aluminum. The unfinished surface may show marks of molds or rolls and is light gray in color. The newly filed surface is smooth and white. The chips are smooth, have saw edges where cut,

METALS USED IN AIRCRAFT

and can be continuous if desired. There is no spark when applied to the grinding wheel.

Aluminum and its alloys melt faster than steel with no apparent change in color. A stiff black scum usually forms which is quiet. The molten puddle is the same color as the nonheated metal and is fluid. The slag is quiet.

To distinguish between aluminum and aluminum alloys, it is only necessary to dip a small portion of the metal in a 10 per cent solution of caustic soda for a few seconds. The alloy will turn dark and remain dark after an ordinary wash in water. Aluminum also turns dark, but the discoloration may be washed off with water. In case the sheet is Alclad, the surface will test as pure aluminum, but the edge will show a dark area which is permanent.

The flame test may be used to identify magnesium alloys. A small piece of the metal heated to its melting point will ignite and burn with a bright flame. Aluminum or aluminum alloys will form into a small ball of molten metal which is characteristic of these alloys under heat.

Monel Metal. Monel metal is fairly smooth and light gray on the newly broken surface. The unfinished surface is smooth and dark gray. The freshly filed surface is smooth and light gray. The chips have smooth edges which may be continuous if desired. Monel metal chips easily. The spark stream is about 10 in. in length, small in volume, and the sparks form small wavy streaks similar to nickel. The spark stream is orange in color with no sparklers. Monel metal melts more slowly than steel and becomes red before melting. The slag forms a gray scum in considerable amounts and is quiet and hard to break up. The molten puddle is fluid under the slag and is quiet.

Nickel. Nickel is slightly rough and almost white on the newly broken surface. The unfinished surface is smooth and dark gray. The freshly

Fig. 139. Sparks produced from nickel by a grinding wheel. Note small volume and wavy streaks.

filed surface is smooth and a bright silvery white. The chips have smooth edges and can be continuous if desired. Nickel chips easily. The spark stream is about 10 in. in length, small in volume, and the sparks form small wavy streaks with no sparklers. The spark stream is orange in color.

Lead. Lead is light gray and crystalline on the newly broken surface. The unfinished surface is smooth, velvety, and from white to gray in color. The newly filed surface is smooth and white. Any shape of chip may be obtained because of the softness and it may be continuous if desired. Lead is soft enough to cut with a knife. It shows no spark on a grinding wheel and melts fast with no apparent change in color.

Metals. Alloys Used in Aircraft Construction. Aluminum is the only metal used to any extent in its pure state in aircraft construction. Alloys of aluminum, steel, copper, and nickel are the most important used in the construction of aircraft.

The following metals are commonly used to form the alloys of aluminum: iron, magnesium, manganese, copper, silicon, zinc, chromium, nickel, lead, and bismuth.

The metals commonly alloyed with iron are nickel, chromium, molybdenum, silicon, copper, vanadium, tungsten, manganese, magnesium, and aluminum.

The strong alloys of aluminum, such as Duralumin are used in such engine parts as cylinder heads, crank cases, connecting rods, and pistons, and are used in the fabrication of structural parts and the skin of an aircraft.

Commercially, pure aluminum is usually used in the sheet form for parts fabricated by welding. It is also obtainable in seamless tubes, rods, wire, and extrusions.

The alloys, 2S, 3S, and 52S, are used in fuel- and oil-tank construction, air intakes, air scoops, cowlings, fairings, and other parts fabricated by flame welding. The tube form of aluminum alloys is used for oil and fuel lines. These alloys may be identified by color markings painted on the metal or at the end of rods and tubes. The specification number is usually stamped on the metals.

Nickel steel is available in bars or sheets. It may be obtained in the cold-rolled or annealed condition for general machining and forging. It is used for parts subject to high stresses and wear, such as wrist pins, crankshafts, turnbuckles, clevises, cam plates, fittings, and bolts. Most

METALS USED IN AIRCRAFT

of the nickel steels are hard and are worked with difficulty except at high temperatures.

Nickel chromium steel is used where strength, ductility, toughness, and shock-resistant properties are required, such as for crankshafts, drive shafts, gears, and connecting rods.

Chromium molybdenum steel is used for aircraft-engine parts where toughness and strength are desired, such as connecting rods, and may

Fig. 140. A welded steel fuselage showing the engine mount and landing gear. (Courtesy Fairchild Engine and Airplane Corporation)

be obtained in bar, sheet, and tubular form. The sheet and tubular forms are used principally for fuselage and landing-gear struts and other built-up welded structures where heat treatment is impractical. This grade of alloy steel is being used extensively for various airplane structures, because of its quality of retaining most of its normal strength after welding without heat treatment.

Chromium steel is especially noted for its stainless and noncorrosive qualities and is used for aircraft skins and other parts where corrosion-resistant qualities are necessary.

Chromium vanadium steel is strong and tough and is used for wrist pins, pinion gears, and bearing races. Its tensile strength after heat treatment is from 150,000 to 200,000 p.s.i.

Tungsten steel is noted for its hardness and heat-resisting qualities. The low-tungsten grade in the annealed condition is used for permanent magnets. The high-tungsten grade is used principally for exhaust valves. It retains its tensile strength at high temperature.

Silicon-manganese steel is seldom used in airplane construction, although it is suitable as spring steel for use in tail-wheel springs.

Copper is used in its pure form in electrical equipment and wiring, because of its high electrical conductivity and low cost.

Brass is an alloy of copper and zinc. There are a number of different brasses, depending upon the amount of the alloying material added to the copper. Brass is used in bearings, bushings, and fittings where a soft bearing surface is desired. It is also used in brazing.

Bronze is an alloy of copper and tin. The various bronzes depend upon the proportion of copper and tin present in the alloy. Bronze may be modified by the addition of phosphorus, forming phosphor bronze, or aluminum, forming aluminum bronze. The trade names for some of the copper alloys are: Everdur, High Brass, Red Brass, and Muntz Metal. Bronze is used in bronze welding and for bushings, fittings, and bearings.

The alloys of nickel are used in aircraft fabrication in the form of Monel metal, Inconel and stainless steel. Inconel is an alloy containing about 75 per cent nickel, 12 to 15 per cent chromium, and 9 per cent iron. It also has small amounts of carbon, copper, manganese, and silicon. Monel metal contains approximately 75 per cent nickel and 25 per cent copper, with small amounts of iron, manganese, silicon, and copper. It is used for collector rings and exhaust stacks and is furnished in the form of bars and sheets.

Lead and tin are alloyed for soft solders and combined with silver to make silver solders of various degrees of hardness.

Sheet and Plate. Almost all of the metals and their alloys may be obtained in the sheet form. The composition of the sheet may be the same as in tubes, bars, forgings, or castings.

Material up to a thickness of $\frac{1}{8}$ in., which is approximately 11 gauge, is usually classified as sheet. When the rolled material exceeds $\frac{1}{8}$ in. in thickness, it is usually classified as plate. Very thin material is called foil or leaf. The sheets used in aircraft construction are usually designated by their thickness in decimals of an inch, rather than the gauge which is more commonly used in other industries. The thickness of the metal in terms of gauge varies with the different numbering systems used. For example, in the American Wire gauge and Brown and Sharpe systems, 16 gauge has a decimal equivalent of 0.061 in.; while in the

METALS USED IN AIRCRAFT

Fig. 141. Aluminum alloy sheets to be used in airplane construction. (Courtesy Reynolds Metal Company)

Birmingham Wire gauge and Stub systems, 16 gauge has a thickness of 0.065 in. These thicknesses are roughly equivalent to $\frac{1}{16}$ in.

The aluminum alloys, 2S, 3S and 52S, are usually used in the sheet form for nonstructural parts where high strength is not necessary, such as air scoops, cowlings, fairings, and fuel and oil tanks, and parts which are to be fastened by welding. These alloys are softer than other alloys of aluminum and are more easily formed.

The strong alloys of aluminum, 17S–T, 24S–T, and 75S–T, are used in structural parts where strength is necessary, such as spars, ribs, bulkheads, stringers, and the skin. These alloys are hard, stiff, and difficult to form in their heat-treated condition. When considerable forming is to take place, the sheets of these alloys are usually formed in the annealed condition and heat-treated afterward.

Stainless steels, such as 18–8, are used in the sheet form for structural members and fittings where high tensile strength and noncorrosive qualities are required. This material has recently come into considerable use as skin covering for aircraft. Stainless-steel sheets should

AIRCRAFT MAINTENANCE AND SERVICE

Fig. 142. Anodizing a riveted aluminum fuel tank. (Courtesy Aluminum Company of America)

be fastened by means of rivets or spot welding where high-strength joints are required. While stainless steel may be successfully welded with the oxy-acetylene flame, the heat of the operation tends to destroy the heat-treat qualities of the material in the vicinity of the weld.

Copper and copper alloys in the form of sheet are used to some extent in aircraft fabrication where their particular corrosion-resisting qualities are desired. Sheets of nickel alloys, such as Inconel and Monel metal, are used where heat-resistant and noncorrosive qualities are necessary, as in collector rings and exhaust sticks.

Heat Treatment. The physical properties of most metals and alloys can be changed by heat treatment. Heat treatment consists of heating and cooling under various conditions.

Heat treatment not only affects the grain structure of the metal, but also may affect its chemical composition. As a rule, the longer the cooling time of a metal or alloy which has been heated above its critical temperature, the larger the grain structure. A few of the metals become

METALS USED IN AIRCRAFT

softer when cooled rapidly, but most become harder. The critical point is the temperature to which the metal must be heated to allow a change in the crystalline structure. At ordinary temperatures, changes in crystalline structure may take place, but the process is slow. Most critical temperatures require heating to some degree of redness. At this temperature, crystalline structures may change rapidly. Aluminum shows no changes in color when heated.

With proper heat treatment, such characteristics as hardness, toughness, springiness, ductility, and workability may be controlled. Usually the heating is brought about gradually, and the metal is held at the required temperature for varying lengths of time.

Annealing, hardening, tempering, and normalizing are forms of heat treatment.

Special heat treatments, such as carburizing, cyaniding, case hardening, and nitriding, are sometimes used. Heat treatment can be used to relieve stresses and strains within the metal which have been set up by hammering, rolling, working, welding, and other operations to which the material may have been subjected.

Fig. 143. These sheets of aluminum alloy show the distortion brought about by heat treating. (Courtesy Reynolds Metal Company)

AIRCRAFT MAINTENANCE AND SERVICE

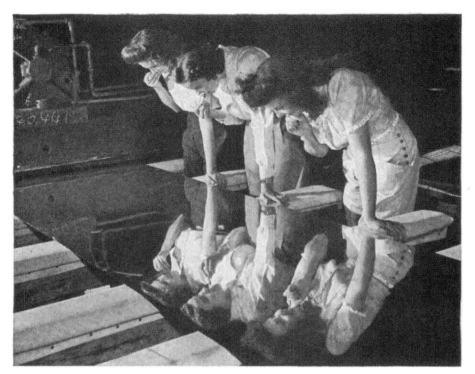

Fig. 144. The above photograph was taken to show the mirror-like finish on aluminum alloy sheets as they come from the rolls. The employees often use these sheets as a mirror. (Courtesy Reynolds Metal Company)

Aluminum alloys which have been cold-worked become strain-hardened. This condition may be removed by annealing.

In the heat treatment of the aluminum alloys, great care is required to obtain the desired results. The alloy must be raised to a temperature which will permit recrystallization in order that the strain-hardened condition be removed, but the temperature must not be high enough to destroy the properties derived from the regular heat treatment. In other words, the strains set up by the cold-working may be removed independently of the hardness and strength of the material due to the heat treatment. Heating to too high a temperature would completely soften the material and require reheat treatment. The heating of these alloys to a temperature between 630° and 650° F. will remove the hardness brought about by cold-working.

The annealing and heat treatment of alloys such as stainless steel are quite different from those processes used for aluminum alloys. Annealing and heat-treatment operations accomplish one or more of

METALS USED IN AIRCRAFT

the following results: (1) restore corrosion-resisting qualities which may have been destroyed by working at high temperatures or by welding, (2) relieve stresses set up in the material by cold-working, (3) obtain special physical properties in alloys which respond readily to heat treatment.

XVI ALLOYS USED IN AIRCRAFT

Iron and Iron Alloys. Steel is iron which contains a varying percentage of carbon. When the carbon content is less than $\frac{1}{10}$ of 1 per cent, the product is considered as iron.

The following is a general classification of the properties of steel, based on the amount of carbon present.

	PER CENT
Dead soft steel	0.10 maximum carbon
Soft steel	0.20 maximum carbon
Mild steel	0.15 to 0.25 carbon
Medium steel	0.25 to 0.45 carbon
Hard steel	0.45 to 0.85 carbon
Spring steel	0.85 to 1.15 carbon

There are many special trade names for steel, such as tool steel, spring steel, firebox steel, high-speed steel, boiler steel, and automobile steel.

Wrought iron is an alloy of iron and carbon in which the carbon content is less than 0.10 per cent and the manganese content is less than 0.07 per cent. Wrought iron is tough, ductile, and soft, and is easily worked.

When iron has a carbon content of less than approximately 0.06 per cent, it is called low-carbon iron. Low-carbon steels, or mild steels, have a range of carbon content from 0.10 to 0.30 per cent. Low-carbon steels are easily worked.

Some of the more common alloys of steel are molybdenum steel, chromium steel, chromium molybdenum steel, chromium vanadium steel, manganese steel, manganese vanadium steel, nickel steel, nickel chromium steel, chromium molybdenum vanadium steel, copper nickel steel, and molybdenum nickel steel. Most of these alloys have trade names and are used widely throughout the aircraft industry.

ALLOYS USED IN AIRCRAFT

Fig. 145. Welded steel wing compressing members. (Courtesy Fairchild Engine and Airplane Corporation)

Chromium molybdenum steel is used extensively in aircraft manufacturing and does not require heat treatment after welding. Many parts of light airplanes have a welded chromium molybdenum steel framework. Most light aircraft have welded steel fuselages. Parts such as rudders, fins, stabilizers, landing gear, engine mounts, and fittings are commonly built up by welding.

Stainless steels have come into use in the last few years. Stainless steels are alloys of iron, chromium, carbon, nickel, and other metals. Stainless steels are known to the trade by their number classification, such as 18–8 and 25–20. One of the most common stainless steels is 18–8, which contains approximately 18 per cent chromium, 8 per cent nickel, and from 0.08 to 0.20 per cent carbon alloyed with iron. These alloys are quite difficult to draw and form.

There are a number of "clad" steels in use such as copper-clad, nickel-clad, and stainless-clad. Clad steels are built up by covering a sheet of the metal on one or both sides with a sheet of another material. These

AIRCRAFT MAINTENANCE AND SERVICE

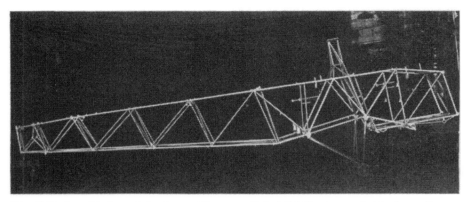

Fig. 146. A welded steel fuselage. (Courtesy Fairchild Engine and Airplane Corporation)

outer layers are usually used to prevent corrosion, although they may add strength and wear-resisting qualities. Galvanized steels are formed by dipping sheets of steel into molten zinc.

Terneplate is sheet iron or steel which has been coated with an alloy of lead and tin. It is commonly used to make patterns or templates in sheet-metal work.

Aluminum and Aluminum Alloys. Strong alloys of aluminum used for aircraft structures depend mainly on the amount of the alloying substances which is combined with the aluminum. Many of the alloys of aluminum depend for their strength and hardness upon proper heat treatment.

The average composition of some of the commercial alloys of aluminum are listed in the following table.

TABLE VII. NOMINAL PERCENTAGE OF ALLOYING ELEMENTS OF ALUMINUM ALLOYS *

ALLOY	SILICON (Si)	COPPER (Cu)	MANGANESE (Mn)	MAGNESIUM (Mg)	CHROMIUM (Cr)
3S	....	....	1.20	....	....
14S	0.80	4.40	0.80	0.40	....
A17S	....	2.50	....	0.30	....
17S	....	4.00	0.50	0.50	....
24S	....	4.50	0.30	1.50	....
25S	0.80	4.50	0.80	....	....
51S	1.00	....	....	0.60	....
52S	....	....	....	2.50	0.25
53S	0.70	....	....	1.30	0.25
61S	0.60	0.25	....	1.00	0.25

* Balance composed of aluminum and nominal impurities.

ALLOYS USED IN AIRCRAFT

Fig. 147. Electric welding of the light steel tubes which help to form the fuselage structure. (Courtesy Consolidated Vultee Aircraft Corporation, Stinson Division)

Annealing is commonly used to soften aluminum alloys. After aluminum has been cold-worked, it is usually heat-treated before being used in the fabrication of aircraft.

Temper Designations. 2S, 3S, and 52S aluminum alloys are termed common alloys, while 14S, A17S, 24S, 52S, and 61S are termed strong alloys. All of these alloys can be produced in the softest possible state when annealed. The S designates a wrought alloy, one that is produced by working, rolling, or extruding. Alloys that have been annealed are designated by an O; for example, 2S–O and 24S–O.

2S, 3S, and 52S alloys when produced as hard as possible are designated by an H; for example, 2S–H, 3S–H, and 52S–H. These alloys are also produced in tempers in between soft and hard, such as quarter-hard (2S–¼H) and half-hard (2S–½H).

The strong alloys can be heat-treated to produce much harder and stronger materials. 17S and 24S alloys are designated 17S–T and 24S–T when heat-treated. 53S and 61S can be produced in two heat-treated

AIRCRAFT MAINTENANCE AND SERVICE

Fig. 148. Aluminum alloy sheets to be used for aircraft skins being cooled by water after annealing. (Courtesy Douglas Aircraft Company, Incorporated)

conditions, designated W and T, such as 53S–W and 53S–T. The W condition is not as hard as the T condition. The T condition is produced by heating material in the W condition to temperatures ranging from 320° F. to 350° F. This second heating operation is termed aging or precipitation heat treatment. When the alloy is heated and quenched to produce the W condition, it is called solution heat treatment. The fabricated condition is designated by an F; for example, 2S–F. The F is used to designate material in the "as extruded," "as rolled," or "as drawn" state. Heat-treated alloys may be further strengthened and hardened after heat treatment by rolling. For example, 17S–T after rolling would have the designation 17S–RT.

The following gives a summary of the conditions, or tempers, in approximate order of increasing hardness or strength:

O	Annealed (soft)	
F	Fabricated (as extruded, rolled, or drawn)	
¼H	One-quarter hard	
½H	One-half hard	
¾H	Three-quarters hard	
H	Full hard	*Courtesy Reynolds Metal Company*

ALLOYS USED IN AIRCRAFT

- T   Solution heat-treated (but not aged)
- W   Solution heat-treated and precipitation heat-treated (aged)
- R   Rolled after heat treatment

Courtesy Reynolds Metal Company

Hardness and strength vary with the method of forming, as shown in Table VIII which gives the various uses of the more common aluminum alloys and their mechanical properties and chemical composition.

TABLE VIII. TYPICAL PHYSICAL PROPERTIES OF PRINCIPAL ALUMINUM ALLOYS FOR AIRCRAFT

ALLOY AND TEMPER	COMMODITY	TENSILE STRENGTH (LBS. PER SQ. IN.) *	YIELD STRENGTH (LBS. PER SQ. IN.) *
17S–O	Sheet	27,000	12,000
	Tubing	28,000	12,000
	Extrusions	30,000	12,000
	Rod	30,000	12,000
Pure-clad 17S–O	Sheet	26,000	11,000
17S–T	Sheet	63,000	41,000
	Tubing	64,000	42,000
	Extrusions	60,000	40,000
	Rod	62,000	35,000
Pure-clad 17S–T	Sheet	58,000	36,000
17S–RT	Sheet	68,000	51,000
24S–O	Sheet	27,000	12,000
	Tubing	28,000	12,000
	Extrusions	30,000	12,000
	Rod	30,000	12,000
Pure-clad 24S–O	Sheet	25,000	11,000
24S–T	Sheet	68,000	46,000
	Tubing	67,000	46,000
	Extrusions	68,000	50,000
	Rod	68,000	43,000
Pure-clad 24S–T	Sheet	63,000	42,000
24S–RT	Sheet	72,000	55,000
52S–O	Sheet	28,000	14,000
	Tubing	30,000	14,000

* Approximate values

Courtesy Reynolds Metal Company

Magnesium and Magnesium Alloys. Because of its lightness, magnesium is coming into wide use in aircraft structures. Its low specific gravity of 1.74 makes it the lightest of the structural metals. Magnesium is a silvery white metal which, when alloyed with aluminum, zinc, manganese and other metals, produces an alloy which has the highest strength-weight ratio of any of the commonly used structural metals. This metal weighs only about one-fifth as much as steel and about two-thirds as much as aluminum. The melting point is about 1204° F. Magnesium tends to combine readily with oxygen, and pure magnesium will burn violently. It may be worked readily by machining and be formed by standard methods. Magnesium does not form in the cold condition as readily as it does at temperatures ranging from 250° F. to 600° F.

Some magnesium alloys are furnished to the industry under the trade name, Dowmetal products. These alloys contain magnesium in percentages varying from 89.9 to 98.5. The magnesium is alloyed with manganese, aluminum, silicon, zinc, and tin. These alloys are coming into use rapidly because of their high strength-to-weight ratio. The magnesium alloys are used for such parts as crankcases and structural parts of aircraft.

Magnesium alloys may be fastened by riveting, spot welding, electric or gas welding. It is necessary, when welding magnesium by either the electric arc or the gas flame, to protect the hot metal from the oxygen of the air by either a non-inflammable gas such as helium, or by a suitable flux which will keep the metal covered.

Table IX gives the typical analysis of commercially pure magnesium.

TABLE IX. TYPICAL ANALYSIS OF COMMERCIAL MAGNESIUM

ELEMENT	PER CENT
Aluminum	0.003
Copper	0.003
Iron	0.03
Manganese	0.08
Nickel	0.001
Silicon	0.005
Magnesium (by difference)	99.878

Courtesy Reynolds Metal Company

Table X shows the composition and use of magnesium alloys with their letter designation.

Magnesium alloys do not corrode in dry air. Ordinary air, such as encountered in inland districts, does not cause excessive corrosion.

ALLOYS USED IN AIRCRAFT

TABLE X. COMPOSITIONS AND USES OF MAGNESIUM ALLOYS

DOW-METAL ALLOY	NOMINAL COMPOSITION * (Per Cent)			USE
	Al	Mn	Zn	
C	9.0	0.1	2.0	Sand and permanent mold castings. Heat-treatable.
FS-1	3.0	0.3	1.0	Extrusions, plate, sheet, and strip.
G	10.0	0.1	—	Permanent mold castings. Heat-treatable.
H	6.0	0.2	3.0	Sand castings. Heat-treatable.
J-1	6.5	0.2	1.0	Extrusions, forgings, and sheet.
M	—	1.5	—	Extrusions, forgings, plate, sheet, and strip. Used also for special sand castings.
O-1	8.5	0.2	0.5	Extrusions and forgings. Heat-treatable.
R	9.0	0.2	0.6	Die castings.

* Remainder magnesium.

Courtesy Reynolds Metal Company

The metal does become covered with a thin gray film consisting of the oxide and carbonate of magnesium. The presence of salt in the air, as encountered along the sea coast, or salt water increases corrosion. Magnesium not only corrodes in contact with other metals, but will corrode when in contact with wood in the presence of moisture. Any particle of foreign material which may be embedded in the surface of magnesium alloys during the process of working will cause corrosion. These tiny particles act as minute electric cells. Magnesium alloys should not be allowed to rub against steel, be cleaned with a wire brush, steel wool, or emery cloth, nor be sandblasted with any material. The metal in the area of a gas weld must be thoroughly cleaned of all flux or impurities to prevent corrosion. Magnesium alloys exposed to corrosive conditions should be protected by one or more coats of zinc chromate primer covered with suitable coats of paint.

Standard rivet heads and riveting practices may be used in fastening magnesium alloys. The rivets commonly used are of the following aluminum alloys: 2S, 3S, 53S, 61-T, 56S-O, 56S-$\frac{1}{4}$H and A17S-T. The alloy, 56S, is recommended because there is less possibility of corrosion. Rivets of steel, copper, brass, and other heavy metals should not be used because of the danger of electrolytic corrosion. Magnesium alloys are not used for rivets because of the rapid work-hardening which takes place when a rivet is driven cold. Rivets of aluminum alloys other than 56S should be dipped in a zinc chromate primer to reduce the possibility of corrosion. All riveted assemblies of magnesium must be painted to protect them from corrosion. If it is necessary in an emergency, such as a field repair, to use rivets of steel, copper, or brass, the rivet should be chromium plated and dipped in zinc chro-

AIRCRAFT MAINTENANCE AND SERVICE

Fig. 149. Aircraft skin formed of heat-treated aluminum alloy sheets. (Courtesy Douglas Aircraft Company, Incorporated)

mate primer before driving. When sheets of magnesium alloys are riveted to dissimilar metals, the metals should be separated by a gasket of corrosion-resistant material or a suitable sealing compound. The joint between magnesium sheets which are to be riveted should be painted with two coats of zinc chromate primer before assembly. It is advisable to coat all rivets, even the alloy 56S, with zinc chromate primer before driving.

After forming parts from magnesium alloys, the parts should be submerged in a solvent to remove all impurities from the surface of the metal. The first bath should consist of an alkaline solution which will remove all traces of grease and oil. The material should then be treated to an anodizing solution, such as a chromic acid bath, rinsed in cold running water, and protective primers and paint applied.

XV FABRIC–COVERED CONSTRUCTION

Construction of a Fabric-Covered Wing. The bay of a wing is that portion of the wing which lies between successive compression members.

Each bay consists of the following parts: front spar, rear spar, compression members, glue blocks, wood braces, fittings, wing covering, form ribs, false ribs, leading-edge strip, leading-edge cover, trailing edge, and drag and antidrag wires.

The root bay constructed in the following problems includes different types of construction which may be found in any standard wing.

A number of different ribs will be constructed so that the wood-

Fig. 150. Electric welding of the heavy tubes forming the main fuselage structure on a light, post-war airplane. (Courtesy Consolidated Vultee Aircraft Corporation, Stinson Division)

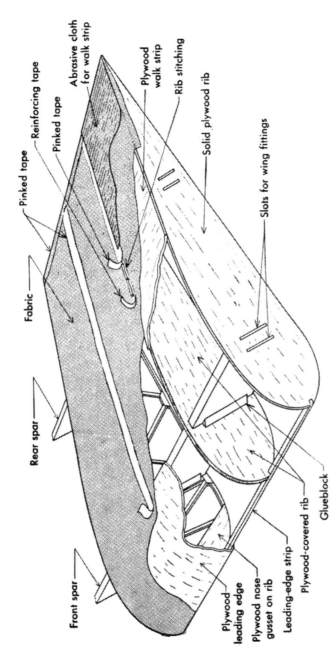

Fig. 151. Sketch showing the construction and parts of a root bay of a fabric-covered wooden wing. (Not a scale drawing.) Root of right wing.

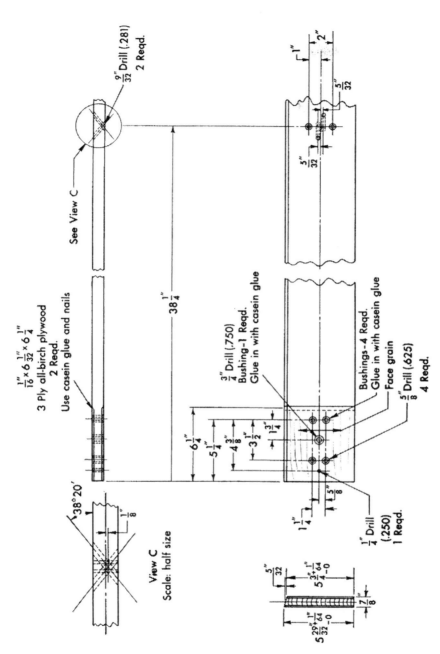

Fig. 152. The construction of the root portion of the left front spar.

worker will become familiar with various types of ribs. All construction is standard.

This bay will be 38¼ in. in length from the root of the spar to the center of the fittings to which the drag and antidrag wires are fastened. The first rib will consist of a root or walk rib. This rib is also called a box rib. It will be covered with plywood and have a walk strip covered with abrasive cloth. The walk or box rib consists of the root compression rib and a reinforced rib covered with a plywood walk.

In addition to the root or box rib, there is a form rib, a reinforced compression rib, and two false ribs. The compression rib must support the strain placed on it by the drag and antidrag wires. The false ribs are the short ribs at the leading edge of the wing which maintain the shape of the leading edge between the main or form ribs.

Root-Bay Section of the Front Spar. For the front spar, a spruce plank 6½ in. wide, 1 in. thick, and not less than 48 in. long will be needed. Figure 153 gives the dimensions of the cross section of the spar, and these dimensions should be followed exactly. The piece of spruce plank should comply with the specifications given for spar stock. Plane one side and edge of the plank until smooth and at right angles. Care should be taken that this edge is exactly straight, as it will be the bottom edge of the spar. Carefully square one end of the plank which will be the root end of the spar. The straightness may be tested by a steel straightedge. The right angle may be tested with a try square. The second side of the spar should be dressed down until the spar is ⅞ in. thick. Bevel the top of the spar. Care should be taken that this bevel is exact. The depth of the front face of the spar is 5¾ in., and the depth of the rear face is $5^{29}/_{32}$ in., as shown in Figure 153.

It is good practice, whenever fittings are to be bolted to spars or other structural members, to reinforce these points with plywood. These pieces, as shown in Figure 152, extend 6¼ in. from the root of the spar and from the bottom edge of the spar to the beveled edge. The pieces should be beveled to match the bevel on the top edges of the spar after they are in place. To install these plywood reinforcings, a line should be drawn across each face of the spar parallel to the root end of the spar, exactly 6⅛ in. from the end of the spar. A saw cut 1/16 in. deep should be made, and 1/16 in. of the spar material should be carefully removed by the use of a broad chisel or a block plane. When this material has been removed to the proper depth, the square corner left by the saw cut should be beveled. The plywood pieces

FABRIC–COVERED CONSTRUCTION

which are to fit on each side of the spar should have one edge beveled to the same slope as that part of the spar just beveled. This joint should be as nearly perfect as possible. The plywood pieces are fastened in place by the use of casein glue and nails. These nails should be aircraft

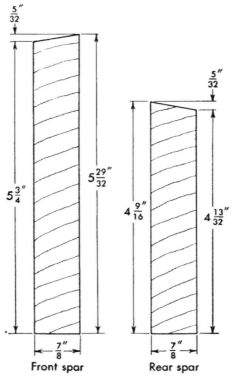

Fig. 153. Cross section of a front and a rear spar.

nails and enter the spar at least 3 times the thickness of the plywood. Plywood should be prepared for both sides of the spar and glued at the same time.

Gluing Plywood Reinforcements. Proper gluing blocks should be applied. C-clamps should be used to exert pressure. The plywood should be lightly sanded to remove the finish and give a better gluing surface. The plywood on the rear of the spar should be placed in position, glue applied, and nailed with staggered rows of nails approximately 1 in. apart. Plywood is applied to the front face of the spar in the same manner. Before placing the gluing blocks on each side of the spar, cover the plywood and the glued edges with a sheet of paper

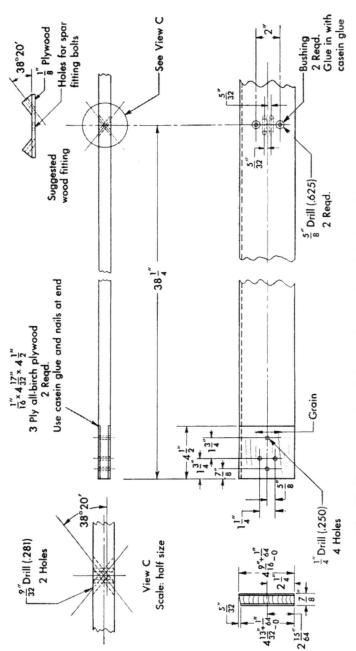

Fig. 154. The construction of the root-bay portion of the lower left rear spar.

FABRIC–COVERED CONSTRUCTION

Fig. 155. Wing construction showing a molded plywood leading edge, gussets, and rib webs. (Courtesy Universal Products Corporation)

to prevent the glue sticking to the glue block and ruining the plywood when the glue block is removed. A pressure of approximately 150 lb. per sq. in. should be applied. After about 10 min. the clamps should be retightened because the glue being forced into the wood relieves some of the pressure. Allow glue to set 6 to 8 hours.

Drilling Holes for Fittings. The drilling of holes for aircraft fittings requires a high degree of skill. The proper method for drilling the fitting holes at the root of the spar is by use of a hardwood jig.

Root-Bay Section of Rear Spar. The rear spar is constructed in the same manner as the front spar except for the difference in the dimensions. The rear spar is $7/8$ in. thick, the front face is $4\,9/16$ in. in depth, and the rear face is $4\,13/32$ in. in depth. Care should be taken that the bevel on the rear spar is made in the opposite direction from that on the front spar. The narrow face of the front spar is to the front, and the narrow face of the rear spar is to the rear.

Rib Jig. A jig must be prepared for fabricating the ribs in this problem. Select a piece of lumber, preferably hardwood, of the following dimensions: 14 in. wide, 72 in. long, and approximately 1 in. thick. This piece will be used as the base of the rib jig. Draw on this board a straight line 1 in. from the edge of the board. This will be the base or reference line from which the rib will be laid out. At a point 4 in.

from the end of the board, where the nose of the rib will be laid out, erect a perpendicular line to the base or reference line; and 58½ in. from this line erect another perpendicular to the reference line. Along the reference line lay off the dimensions given in Figure 156, the first distance being $31/32$ in., which is the radius of the nose of the rib, then 2½ in., and proceed in this manner until all of the dimensions given along the reference line are clearly marked. These points are called "stations." Note that each dimension given is the over-all dimension from the nose of the rib, the total length of the rib being 58½ in.

Erect a perpendicular at each station on the reference line. On these perpendiculars lay off the dimensions, beginning with the $31/32$ in. station. On the perpendicular at this station, lay off two dimensions, one 3⅛ in. from the reference line, and the other 5⅝ in. from the reference line. On the perpendicular at the 2½ in. station, lay off $2^{19}/_{32}$ in. and $6^{9}/_{16}$ in.; at the $5^{1}/_{16}$ in. station lay off 2¼ in. and $7^{17}/_{32}$ in. on the same perpendicular. At the 9 in. station lay off 2 in. and $8^{7}/_{16}$ in. on the same perpendicular. The remainder of the length of the bottom of the rib is a straight line which will be 2 in. from the reference line. From this point to the trailing edge of the rib only one dimension will be laid off on each perpendicular. At the nose of the rib on the perpendicular erected at the $31/32$ in. station, a point $4^{9}/_{32}$ in. from the base line is marked, which is the center of the $31/32$ in. radius, to form the nose of the rib.

After all dimensions given are marked on the jig base with a flexible piece of wood or plastic, the points outlining the top and bottom of the rib are drawn in as shown in Figure 156. After the rib has been outlined, suitable blocks are prepared from either ⅜ in. hardwood or ⅜ in. plywood and placed around the outline of the rib, as shown in Figure 156. These blocks should be glued and nailed or screwed into place. The block forming the outline of the nose rib should be carefully prepared and glued and screwed into place to form the nose portion of the jig. After the outside blocks are in place and allowed to dry, the measurements of the various stations should be carefully checked. Blocks around the cambered portion of the rib should have the edge next to the line shaped just to touch the line throughout the length of the block.

Prepare a nose block and place in the jig. The block should fit snugly against the nose-forming block and be nailed temporarily into place.

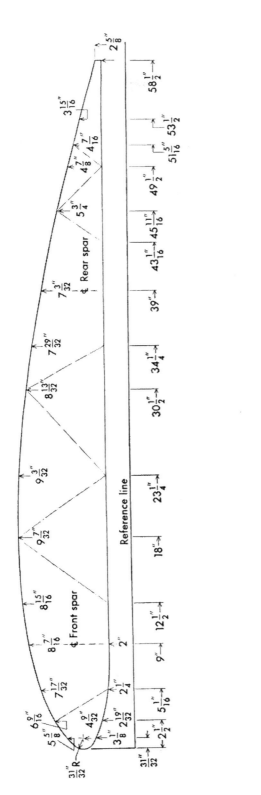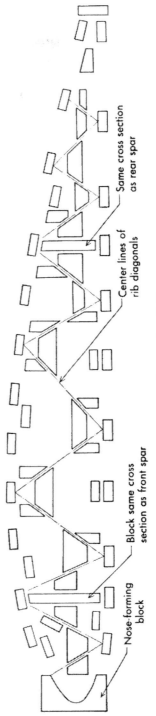

Fig. 156. A layout of a rib jig.

Cap strips for both the top and bottom of the rib should be prepared. This strip should be ⅜ in. square and of straight-grained spruce such as will be used to build the rib. These strips should be slightly longer than the top and bottom cap strips of the ribs. Place these strips in their proper place as though they were top and bottom cap strips, holding them in place by temporary nailing. These strips should fit closely against the blocks forming the outer outline of the top and bottom cap strip. Place the blocks inside the cap strip, as shown in Figure 156, forcing them snugly against the temporary cap strips, and nail and glue. After these blocks are in place, prepare two blocks which are the same in size as the cross section of the front and rear spar. Fasten these blocks in place with glue and screws.

Rib Cross Members. Prepare rib cross members. Place one on each side of the blocks representing the spars. These pieces will be the same as the verticals in the rib on each side of the spar. Place diagonals of ⅜ in. square spruce and fasten temporarily along the center lines as shown in Figure 156. A temporary rib is now in place in the jig. Blocks should be prepared as shown in Figure 156 to fit snugly against these diagonals. These blocks are to be glued and nailed or screwed into place. The jig is complete and should be allowed to dry thoroughly, and the pieces forming the rib removed. The entire jig should be coated with hot paraffin to prevent the rib from sticking to the jig. All dimensions should be carefully checked to see that they agree with the drawing.

Construction of Rib. The different types of ribs will all use the same jig for their main framework. Before placing the cap strips in the jig, prepare a nose block of ⅜ in. basswood as shown in Figure 157. Use ⅜ in. square spruce throughout for cap strips and diagonals for all ribs. The top and bottom cap strips are first placed in position in the jig. Gussets are prepared of 3/32 in. plywood, as shown in Figure 157. Preferably, this plywood should be mahogany with a cottonwood or basswood core. Be sure that all pieces fit snugly in the jig. Coat the gussets with prepared casein glue and nail into place, using flat-head aircraft nails which penetrate the rib material 3 times the thickness of the gusset. A plywood piece for the nose of the rib should be prepared of the same material as the gussets and glued and nailed into place. Allow the rib to dry not less than 6 hr. Remove the rib from the jig, turn it over on a flat surface, and nail and glue duplicate gussets

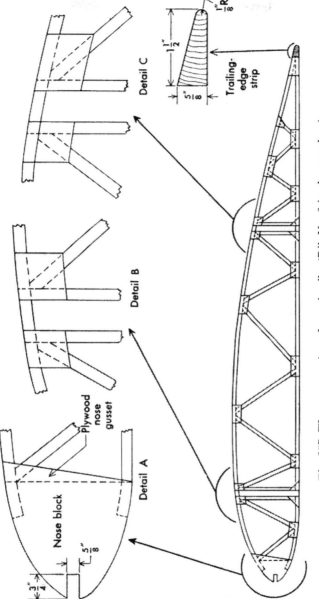

Fig. 157. The construction of a main rib. (Rib No. 3 in the root bay.)

207

AIRCRAFT MAINTENANCE AND SERVICE

Fig. 158. Nail-gluing ribs showing the use of a jig. (Courtesy Forest Products Laboratory)

on the opposite side of the rib. After drying, the rib is complete. This is a regular form rib.

Root Rib. The root rib is formed of solid $3/8$ in. plywood. Any suitable plywood may be used, but should not be made up of less than five plies. This rib is solid plywood, notched at the nose to receive the leading-edge strip. This rib is not cut out to receive the root of the spar, but would normally have openings cut through it for the spar fittings. The rib is fastened to the end of the spars with glue blocks as shown in Figure 159. This root rib acts as a compression rib.

Reinforced Rib. Place in the jig a top and bottom cap strip and nose piece. The verticals on each side of the spars are placed the same as for a main rib. The other cross pieces are not used in this rib. This rib will be covered on both sides with $1/16$ in. plywood. The plywood pieces to cover each side of this rib must have openings cut the same size as the cross section of the front and rear spars, as shown in Fig-

FABRIC–COVERED CONSTRUCTION

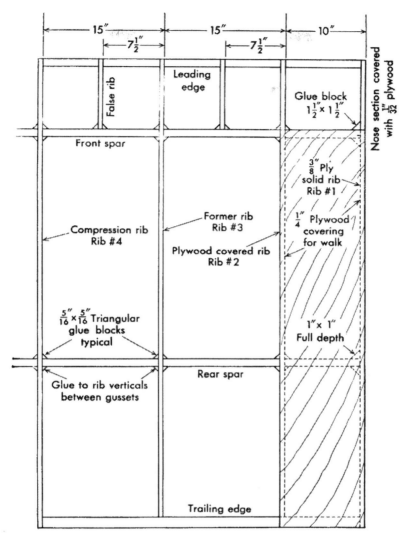

Fig. 159. A plan drawing showing the main structure of a root bay.

ure 153. All portions of the rib in the jig are coated with casein glue, and the plywood webs are placed in position and nailed with the same-sized nails used for the gussets in the main rib. Allow the plywood to dry not less than 6 hr., remove the rib from the jig, placing it on a flat surface, plywood side down, and place the stiffeners in position as shown in Figure 160. The stiffeners should be coated with glue on both sides. The entire exposed portion of the rib to which the plywood will be fastened is also coated with glue. Place the plywood in position, as shown in Figure 160, and nail. Allow the rib to dry not less than 6 hr.

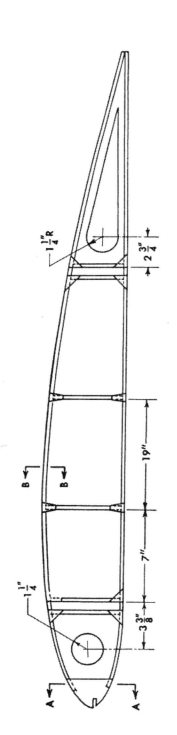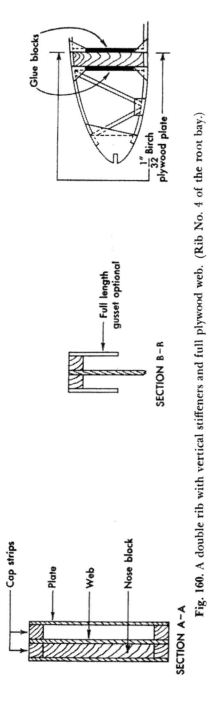

Fig. 160. A double rib with vertical stiffeners and full plywood web. (Rib No. 4 of the root bay.)

FABRIC–COVERED CONSTRUCTION

Compression Rib. This compression rib will have a plywood web and a double top and bottom cap strip with reinforcing verticals. This rib is formed by using a regular top and bottom cap strip, a nose piece, and four verticals. A second half of the rib is built up like the first half, except that the gussets are placed on this half while in the jig. After removing from the jig, the side of the rib opposite the gussets is covered with casein glue and nailed to the plywood web on the first half, using nails which will almost penetrate both portions of the rib. Plywood gussets are now nailed and glued to the opposite side of the rib.

Nose or False Ribs. False ribs are short ribs which help to support the leading edge of the wing and extend only from the front spar forward. False ribs may be constructed by placing top and bottom cap strips in the jig which extend from the nose approximately 2 in. back of the front spar. The rib is then formed in the same way as a regular rib, having the two verticals between which the front spar passes and a nose portion identical to the main ribs.

Leading-Edge Strip. The leading-edge strip is constructed of spruce. This strip is $40\frac{3}{8}$ in. in length, $\frac{1}{2}$ in. thick, and $\frac{3}{4}$ in. wide. The front edge is rounded off, having a radius of $\frac{3}{32}$ in.

Trailing-Edge Strip. This strip is constructed of straight-grained spruce $40\frac{3}{8}$ in. long with a cross section as shown in Figure 157.

Wing Assembly. With all fittings, braces, and glue blocks complete, the root bay of the wing should be assembled as follows. The two spars are placed on supports, and ribs No. 2, 3, and 4, and the two false ribs are put into place. The fittings and wires are now assembled and tightened just enough to hold the spar firmly. The spars should be exactly parallel to each other and the ribs at right angles to the spars. The distances used to line up the assembly may be measured with a trammel gauge. When properly aligned, the glue blocks, braces, etc., are glued and nailed into place. The rib verticals on each side of the spar should be glued and nailed in place. After all the parts are glued and fastened, the contour of the nose portion of the ribs should be checked by using a straightedge. The nose strip which has been prepared should now be fastened into place by the use of casein glue and nails.

Leading-Edge Cover. The leading edge of the wing is covered with $\frac{3}{32}$ in. plywood, preferably mahogany with a cottonwood core. The plywood extends from the rear edge of the top of the front spar around the nose to the rear edge of the bottom of the front spar. The proper-sized sheet of plywood should be prepared. It may be necessary to dampen slightly the outside of the plywood by covering it with wet rags or burlap for approximately $\frac{1}{2}$ hr. This plywood should be glued and nailed into place. Proper nailing blocks between the ribs should already be in place on the top and bottom of the spar. The plywood should be first nailed and glued to the nailing block along the rear edge of the top spar. The edges of the nose portion of the ribs and the blocks on the rear spar should be covered with glue, and the plywood bent around the leading edge and held in place. The plywood is now nailed around the leading edge of the rib, using nails which penetrate the rib not less than 3 times the thickness of the plywood, and similarly nailed along the rear edge of the bottom of the front spar. The nails should be approximately 1 in. apart. The leading edge should be allowed to dry not less than 6 hr. before further work is done on this portion of the wing.

Wing Walk. The wing walk, which covers the root of the wing, is the portion upon which a person steps when entering and leaving the cockpit. This walk extends from the front spar to the trailing edge, and from the root rib to rib No. 2. A strip of $\frac{1}{4}$ in. five-ply mahogany-faced plywood should be glued and nailed to ribs No. 1 and No. 2, as shown in Figure 159.

Wing Covering. The root bay should be covered with standard airplane fabric. It will be necessary to use two strips. The seam should run from front to back. At the root end of the wing bay, 2 in. should be allowed for lapping. At the outer end of the bay, sufficient fabric should be allowed so that it may be fastened together. The fabric beginning at the trailing edge should be carried around the leading edge and back to the trailing edge, allowing sufficient fabric for folding and sewing, as shown in Figure 161. Fabric may be tacked temporarily to the trailing edge while being sewed. The seam should be fastened at intervals of not more than 6 in. Stitches should be not more than $\frac{1}{4}$ in. apart. The thread used for sewing the seam on the trailing edge should be No. 8, four-ply cotton, having a tensile strength of not less than 14 lb. This thread should be lightly waxed with beeswax before using. The fabric should be fastened at the root end of the bay by tacking into place and should be doped to the root rib. To lace the fabric to the ribs, a strip of reinforcing tape $\frac{1}{2}$ in. wide should be placed over each rib on top of the fabric. The cord used for lacing fabric to the ribs may be cotton or linen, having a tensile strength of 80 lb., and should be lightly waxed before using. To lace the fabric

FABRIC–COVERED CONSTRUCTION

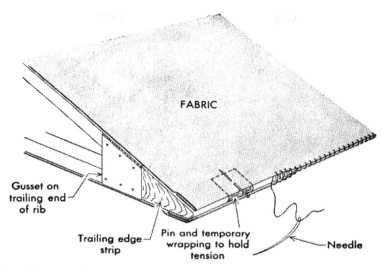

Fig. 161. A method of sewing the fabric at the trailing edge of a wing.

to the rib, a lacing needle of sufficient length to go completely through the wing at its greatest depth is used. The needle carrying the lacing cord is passed downward through the wing as close to the rib as possible, then back through the wing from the bottom to the top on the opposite side of the rib as close to the rib as possible. The cord is then fastened by means of a seine knot. This process is repeated the length of the rib, beginning 4 in. from the front spar. The lacing intervals should be 4 in., each lacing fastened with a seine knot, and the thread carried from one lacing to the next. The wing is now ready for doping and finishing.

Drain grommets should be placed under the trailing edge to allow drainage of any moisture which may accumulate within the wing. Pinked tape of the same material as the wing covering should be used to cover all seams, trailing edges, and ribs. This tape may have either a pinked, scalloped, or frayed edge.

Wing-Walk Tread. The walk rib constructed in the root bay should be covered with an abrasive cloth to prevent slipping when a person steps on the wing, and to protect the fabric from wear. This abrasive cloth is covered with small-sized grit and can be obtained from any aircraft supply house. The material should be cut to fit along the edge of the panel and extend outward to the edge of the wing walk. Pin the material from about 2 in. below the leading edge of the wing over

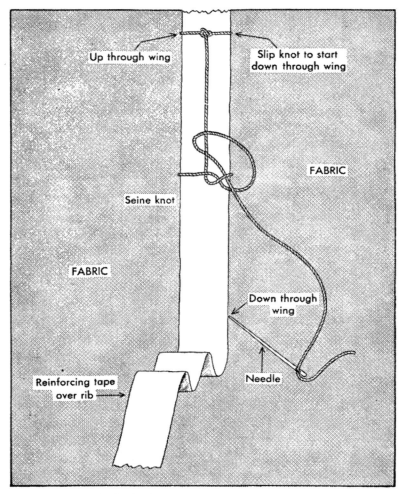

Fig. 162. A method of stitching the fabric to the ribs and seine knot used to fasten each stitch.

the top of the wing and allow it to lap about ½ in. over the trailing edge. With a pencil, mark a guide line across the wing where the outer edge of the wing-walk tread is to be placed. Apply a strip of masking tape along this line on the wing top-side, away from the wing walk. With fine sandpaper, thoroughly sand the fabric which already covers the walk rib. Apply a coat of elastic cement to the underside of the abrasive cloth and to the area of the wing to be covered. After the first coat has dried, spread a second coat of the cement on the abrasive cloth. Allow this coat to set for 5 or 10 min. until it is no longer sticky. Apply a small amount of thinner with a brush to the surface of the cloth covered with

FABRIC–COVERED CONSTRUCTION

cement and immediately apply the cloth to the area to be covered. Apply it first over the leading edge. Use a hand roller and roll the cloth firmly into place. The trailing edge should be fastened in place by a thin metal strip which prevents damage to the thin trailing edge of the wing.

Construction of a Fabric-Covered Aileron. This construction does not include the entire aileron. A portion of an aileron $28 5/8$ in. long will be constructed and finished on each end as though it were a complete aileron. The ribs in the aileron will not run directly from front to back, but will be diagonally placed to form triangular trusses as shown in Figure 163. The aileron spar is of clear spruce, $3/8$ in. thick, $3 3/4$ in. on the front face, and $3 21/32$ in. on the rear face. Two jigs should be constructed to form the end ribs and the angle ribs. The two end ribs are reinforced with $3/64$ in. plywood gussets as shown in Figure 164. One rib will have the gussets on one side and the other rib will have them on the opposite side. The angle ribs will consist of two righthand ribs and two lefthand ribs as shown in Figure 165.

Aileron Spar. The aileron spar will be constructed with the dimensions shown in Figure 164.

Ribs. The lefthand end rib should be constructed and the plywood reinforcing nailed in place while the rib is still in the form. After removing from the jig, the rib is turned over on a flat surface and two small gussets nailed in place. The righthand rib should have the small gussets nailed in place while the rib is in the form, and, after removing from the form, the reinforcing gusset is nailed and glued into place on the opposite side.

Angle Ribs. The reinforcing gussets on the angle ribs are not the same on both sides. Two righthand and two lefthand angle ribs should be constructed. The angle ribs should be $12 1/4$ in. long, and the ends beveled as shown in Figure 165.

Trailing Edge. The trailing edge is constructed as shown in Figure 164.

Assembly. The aileron is assembled by nailing and gluing the ribs to the aileron spar. Then the trailing edge is glued and nailed into place, fastening with $3/64$ in. plywood gussets which have been preformed, extending from the root of the rib on top, around the trailing edge to the end of the rib on the bottom, as shown in Figure 163. The aileron is then covered with fabric and finished in an approved manner.

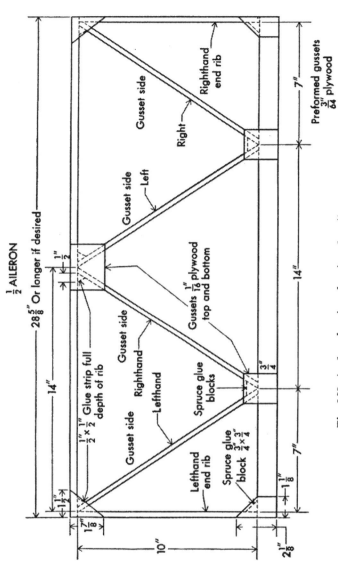

Fig. 163. A plan drawing showing the aileron structure.

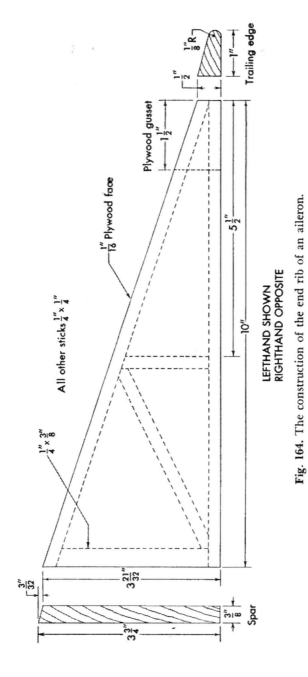

Fig. 164. The construction of the end rib of an aileron.

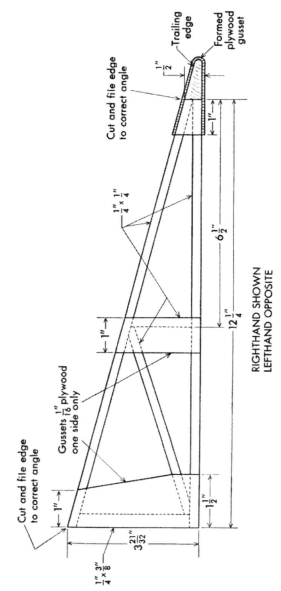

Fig. 165. The construction of an aileron angle rib.

218

FABRIC–COVERED CONSTRUCTION

Construction of a Fabric-Covered Rudder. The control surfaces on an airplane are usually of lighter construction than the stabilizing surfaces. The construction of this rudder is typical of those used even when the rest of the airplane is covered with plywood.

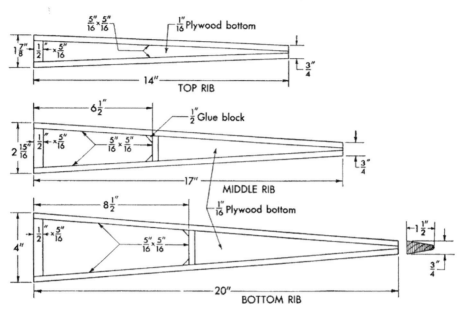

Fig. 166. The construction of a fabric-covered rudder.

Rudder Post. The rudder post should be of clear spruce and be 37½ in. long, 4 in. wide at one end, ⅝ in. wide at the other end, and ⅜ in. thick, as shown in Figure 166.

Trailing Edge. The top and trailing edge of the rudder is of laminated clear white ash ⅛ in. thick and ¾ in. wide. Twelve strips will be needed. At the top edge, this lamination is straight for 6¼ in., then curves with an outside radius of 7½ in. through approximately 75°, and then is straight to the point of junction with the bottom rib of the rudder. A jig will be necessary. After the trailing edge has dried not less than 24 hr., it should be shaped as shown in Figure 166.

Ribs. There are three ribs to be constructed of ⁵⁄₁₆ in. by ⁵⁄₁₆ in. clear spruce, as shown in Figure 167. Only two ribs have a reinforcing cross member which is glued in place. A ½ in. glue block is nailed and glued as shown in Figure 167.

Assembly. The rudder is assembled by nailing and gluing the ribs to the rudder post, then the leading edge should be glued and nailed to the other end of the ribs, as shown in Figure 166. The glue blocks should be glued and nailed into place, and gussets applied at each end of the ribs, as shown in Figure 166. After drying 24 hr., the rudder should be covered with fabric and finished in an approved manner.

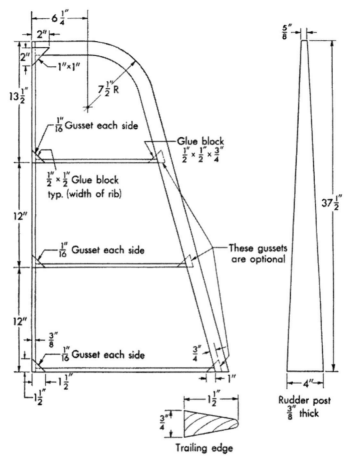

Fig. 167. The rib construction for a rudder.

XVI PLYWOOD CONSTRUCTION

Construction of a Plywood-Covered Wing-Tip Bay. The construction of a wing tip involves several problems which are not encountered in the construction of the root bay or other bays of a wing. When the tip is to be covered with plywood, it requires a somewhat different internal construction than if it were to be covered with fabric.

Spars. The front spar should be 48 in. long, the rear spar 53 in. long. The spars should have the same cross section as the spar in the root bay. The front spar is tapered the last 15½ in. of its length as shown in Figure 168. The rear spar will be tapered for 20½ in., as shown in the same figure. The wing-tip bow should be constructed from laminating ⅛ in. thick strips of white ash.

Wing-Tip Bow. The wing-tip bow should be made of sufficient length to form a 7½ in. scarf joint on both leading and trailing edges. The leading-edge strip should be made of clear white pine and should be ⅝ in. by ¾ in., edge grain on the wide face. A jig should be prepared for the lamination of the wing tip as shown in Figure 169. The curved part of the wing-tip bow, on both the front and rear ends, curves with an outside radius of 20 in. From the point of tangency of these arcs, the remainder of the bow curves with an outside radius of 37 in. Twelve laminated strips should be prepared, ⅛ in. thick, ⅝ in. wide, and approximately 128 in. in length, as the total length of the bow when completed along the outer edge will be approximately 118 in. The strips should be edge grain on the wide face. As shown in Figure 168, semicircular blocks of 1 in. hardwood are sawed, having an outside radius of 18½ in. to form the two ends of the bow. These blocks should be approximately 20 in. in length. A straight block 7½ in. long is cut for the inside of each end of the bow, and the rest of the blocks are cut with an outside radius of 35½ in. to form the center part of the bow. Prepare two outside end blocks for the straight ends of the bow approximately 8 in. long; four blocks, each 6 in. long with

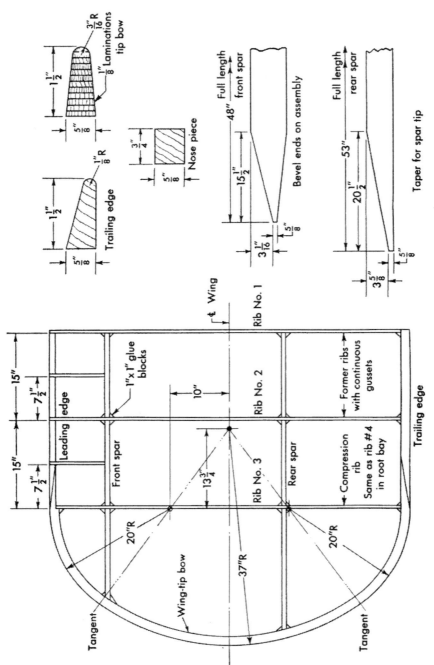

Fig. 168. Plan and detail drawing for construction of plywood covered wing tips.

PLYWOOD CONSTRUCTION

an inside radius of 20 in.; and six blocks 6 in. long with an inside radius of 37 in.

When the ash strips have been prepared, the proper amount of casein glue should be mixed and spread on one side of the first and last strips and on both sides of each of the other strips. The strips

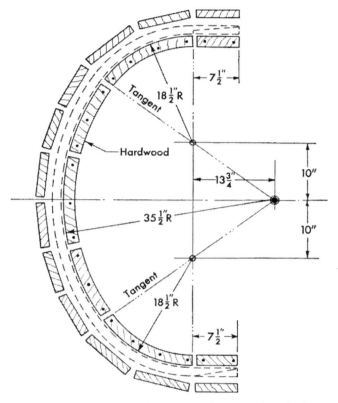

Fig. 169. Construction of jig for laminating wing tip bow.

should be placed together as soon as the glue is spread, and immediately placed in the jig as follows. The straight part of the front end of the bow is clamped into place between the two straight blocks. All of the strips are then bent around the 20 in. radius for approximately one half of the distance when the first block is clamped into place. This procedure is followed all along the bow, clamping each block in place in the proper order. The bow should be allowed to dry in the jig not less than 8 hr., and preferably overnight. After being removed from the jig, the bow should be allowed to dry for an additional 24 hr. before being worked to its proper shape as shown in Figure 168.

Ribs for Wing Tip. This portion of the wing will require three ribs: two form ribs and one compression rib. The compression rib will be the same as rib No. 4 in the root-bay construction. The form ribs will be constructed in a manner similar to those constructed for a fabric-covered wing, except that they are reinforced with plywood gussets extending throughout their entire length. These gussets give the added strength necessary to support the plywood covering. Figure 170 shows

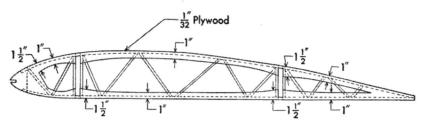

Fig. 170. Continuous gusset for ribs No. 1 and 2 in wing tip.

the method of reinforcing these ribs by a continuous gusset. Two false ribs will be needed which are identical to the false ribs used in the rest of the wing. The drag and antidrag wires and fittings are installed in this portion of the wing in the same manner as that used in the rest of the wing. The framework should be properly rigged before the plywood covering is installed. The leading-edge strip and trailing-edge strip, as shown in Figure 170, should next be constructed. The glue blocks shown in Figure 170 should be prepared, and the entire bay assembled and glued. After drying not less than 24 hr., the plywood covering should be installed.

Wing-Tip Covering. The covering should be of $3/64$ in. mahogany- or birch-faced plywood. It is not practical to cover the entire tip with a single piece of plywood. No horizontal joints are allowed on the leading edge of a wing. A piece of plywood of sufficient size to cover the leading edge between the compression rib and the end of the bay should be formed. This piece will be approximately 30 in. long, and the grain should run around the leading edge from the top of the front spar to the bottom of the rear spar. The edges which will be joined with the other sheets of plywood should be scarfed before forming. When forming, the scarfed edges should be on the inside, except at the outer end, where the scarf should be on the outside. This plywood will be glued by means of a scarf joint to the plywood covering the bay from the front spar to the trailing edge and should overlap. A piece

PLYWOOD CONSTRUCTION

of plywood approximately 30 in. wide and of sufficient length to reach from the front of the front spar on the top of the wing to the trailing edge is scarfed on the front and the outer edge. A similar piece is prepared for the bottom of the tip and scarfed in the same manner. A piece of plywood of sufficient size to cover the tip of the wing from the compression rib to the tip is next cut to shape and scarfed on the bottom side where it will join the leading edge covering and on the bottom side where it joins the covering on the rest of the wing. A similar piece is prepared for the bottom of the wing and scarfed in the same manner. The finish of the surface of the plywood where it will come in contact with the ribs, spars, and wing-tip bow should be scraped or sanded to remove the finish. The pieces covering the portion of the wing back of the front spar toward the root should now be applied, having first had glue spread on that portion of the internal structure, including the ribs, to which it is to be fastened. It is now strip-nailed to rib No. 1, placing the nails approximately 1 in. apart. Next, nail along the rear spar and the top of rib No. 2; then along the front spar. The edge of the scarf portion should reach approximately to the front edge of the

Fig. 171. Nail-strip gluing. (Courtesy Forest Products Laboratory)

block on top of the front spar between the ribs. This edge should be strip-nailed. The rear edge should next be strip-nailed and then the outer edge strip-nailed with a scarf joint reaching the outer edge of the compression rib. The plywood covering on the bottom of the wing

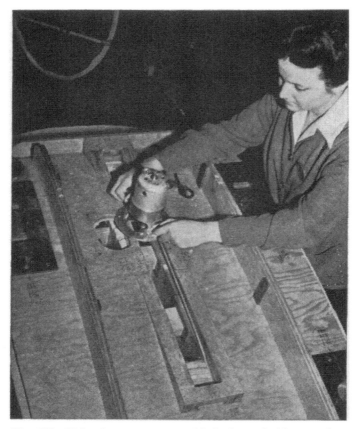

Fig. 172. Girl using router on molded plywood skin on wing. (Courtesy Timm Aircraft Corporation)

opposite this piece should next be applied in the same manner, and these pieces allowed to dry thoroughly. Remove the nailing strips from the front and outside edge of the pieces already in place and apply the leading edge which has been preformed and scarfed on its edges on the inside and on its outer end on the outside. Strip-nail along the leading-edge strip, ribs, and top and bottom of the front spar. After the leading edge has thoroughly dried, apply the pieces of plywood to the tip of the wing. This portion of the wing may be covered with a single piece of plywood, top and bottom, or may be covered with separate

PLYWOOD CONSTRUCTION

pieces, dividing the top and bottom into two or three sections. This piece has the inner edge scarfed on the bottom and is nail-glued over the scarfed edge of the leading edge and the plywood already applied. As soon as it is strip-nailed to the compression rib, the plywood is bent down to the center of the bow by means of hand clamps. These clamps are then placed at the end of each spar and at the center of the 20 in. curved radius, front and rear, and at the front and rear of the

Fig. 173. Center section of molded plywood wing. (Courtesy Timm Aircraft Corporation)

compression rib. Nailing strips are placed between the clamps and nails driven at the end and the middle of each strip, entirely around the wing-tip bow. Nails are then driven between the nails already in place until the interval between nails is approximately 1 in. This method of nailing prevents large wrinkles occurring at any point along the wing tip. If sufficient help is available, all of the plywood pieces may be applied and nail-glued in one operation. No part of the structure should remain unnailed more than 20 min. after the glue is spread, unless clamped firmly in place.

Construction of a Plywood-Covered Horizontal Stabilizer. The construction given here is for the left half of the horizontal stabilizer

AIRCRAFT MAINTENANCE AND SERVICE

which is to be covered with 3/32 in. plywood having a surface veneer of mahogany or birch. The leading edge and the end will be formed of laminated white ash. These laminations should be 2¾ in. wide, ⅛ in. thick, and 74 in. long, edge grain on the wide surface. Twelve strips will be needed. A jig will be necessary, similar to the jig used to form the laminated top and leading edge of the vertical fin. (See Figure 174.)

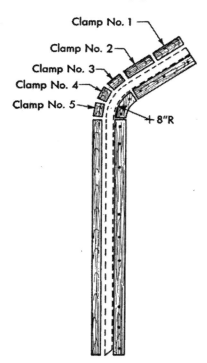

Fig. 174. Construction of jig for vertical fin leading edge.

Leading Edge. The method of fabrication of the leading edge is the same as that used for the construction of the laminated edge of the vertical fin, except that the gluing pressure should be somewhat greater—from 175 to 200 lb. per sq. in. The leading edge is straight for 39 in., then bent on an 8 in. inside radius to approximately 105°, and then straight to where it joins the main spar of the stabilizer. The straight portion at the outer end is first clamped in place, then the portion forming the curve, and finally the straight portion of the stabilizer leading edge. Beginning at a point 39 in. from the fuselage, the leading edge is tapered evenly down to 1 in. wide and 1 in. thick at the point where it meets the spar.

Spar. The spar, which is the rear member of the stabilizer, is made of ½ in. spruce, edge grain on its wide surface. This piece is finished in the same manner as the main spar of the vertical fin. Figure 175 shows the method of construction and the size of materials used. Two strips extend from the fuselage end to the tip of the stabilizer and are constructed and installed in a similar manner to the braces in the fin. Figure 175 shows the method of fastening and the gluing blocks needed.

Covering Stabilizer. The stabilizer is to be covered with 3/64-in. mahogany-faced plywood. The plywood may be applied in one piece by being bent around the leading edge and preformed, or by forming over a suitable form. The straight portion from the fuselage to the beginning of the curved portion should first be glued and nailed in

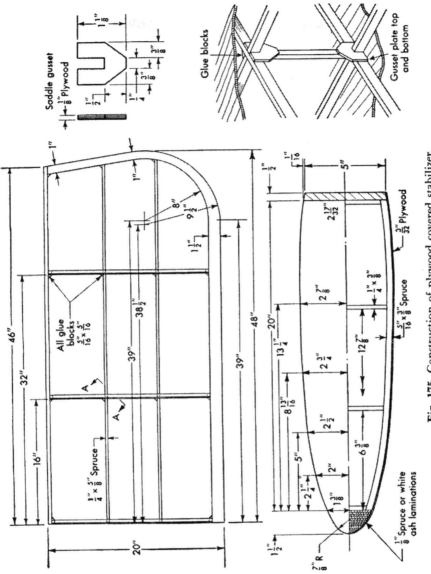

Fig. 175. Construction of plywood covered stabilizer.

place, then nailed along the ribs and rear spar. The tip portion is then glued to the laminated member and held in place by hand clamps while being nailed. The clamps should remain in place until the glue has thoroughly set, not less than 6 or 8 hr.

Construction of a Plywood-Covered Vertical Fin. The vertical fin aids in the directional stability of the airplane. It is usually a fixed surface and not adjustable in flight. In most light airplanes the vertical fin is not adjustable, but is fixed solidly to the fuselage structure. The vertical-fin construction described below is of the fixed type. This fin is of all-wood construction and covered with plywood. The parts to be constructed are the rear vertical member, called the main spar, which is fastened at its lower end to the tailpost of the fuselage; the top and leading edge, which is formed of laminated spruce strips; four horizontal ribs; vertical bracing strips; and glue blocks.

Main Spar. This member should be made from clear, straight-grained spruce having a total over-all length of $73\frac{3}{4}$ in. It is $\frac{3}{8}$ in. thick and $4\frac{3}{8}$ in. wide at its widest part. Select a piece of spruce having not less than 6 annual rings per inch. This piece should be edge grain, $4\frac{1}{2}$ in. wide, $\frac{1}{2}$ in. thick, and approximately 78 in. long. Dress one side and one edge until they are smooth and straight and at right angles to each other. Square one end and dress the piece down until it is exactly $4\frac{3}{8}$ in. wide and $\frac{3}{8}$ in. thick. Mark off from the squared end a distance on each edge of exactly $21\frac{11}{32}$ in. Draw a line across the face exactly $73\frac{3}{4}$ in. from the squared end. Find the center of this line. On each side of the center make a mark $\frac{9}{16}$ in. from the center. Connect this point with the line on the edge $21\frac{11}{32}$ in. from the squared end. Do the same with the other mark. These two lines will taper the vertical member from a point $21\frac{11}{32}$ in. from the squared end to $1\frac{1}{8}$ in. wide at a point $73\frac{3}{4}$ in. from the squared end. With a plane, dress down each side of the main spar to these lines, leaving the excess length at the small end to assist in keeping the end true. When the tapered part is down to the dimensions shown, cut to exactly $73\frac{3}{4}$ in., by sawing off the excess material at the tapered end, and the main spar is complete.

Top and Leading Edge. A jig must be prepared to laminate the top and leading edge. This jig should be similar to the jig used in forming the wing-tip bow. This part of the fin should be formed of $\frac{1}{8}$ in. spruce laminations 2 in. wide. On a bench top or other smooth surface sufficiently large, lay out the inner edge of the laminated member. It will be noted from Figure 176 that this member is straight for a distance

PLYWOOD CONSTRUCTION

of 14⅝ in., then curves on an 8-in. radius to a straight portion extending to a point 45⅜ in. from the main spar. Two-inch blocks should be fastened firmly to the jig base, as shown in Figure 174. The laminations should be of straight-grained spruce, vertical-grained on the wide face. Eleven laminations 80 in. long, ⅛ in. thick, and 2¹⁄₁₆ in. wide will be needed. The extra ¹⁄₁₆ in. should be allowed for smoothing and dressing after removing from the form.

A sufficient number of C-clamps and backing blocks should be on hand before starting the gluing operation. The blocks which hold the curved part should be shaped to the finished curve of the member as shown in Figure 174. Apply casein glue evenly to both sides of each strip with the exception of the two outside strips, piling them one on top of the other as the glue is spread. Beginning at the top, clamp the part that is 14⅜ in. long into place, screwing the clamps up firmly. Carefully bend the remaining strips partly around the curved block, and place the first curved block into position and clamp. Bend still further around the curve, and place the second curved block into position and clamp. Be sure that the edges of all of the strips are even. Finish bending around the curve, clamping the curved blocks into place. Now apply the long, straight block, tightening all of the clamps firmly. After about 20 min., go over all the clamps and retighten them to make sure that sufficient pressure is applied. The pressure applied should be approximately 150 lb. per sq. in. Allow the member to dry not less than 8 hr. before removing from the form. It is better to leave the member in the form 24 hr., if possible. Before dressing down to final form, the member should be allowed to dry at room temperature for 2 or 3 days to allow even distribution of moisture. The fabricated member should then be dressed down to the form shown on the cross-sectional drawings, Figure 176. Note that the edges are tapered from the bottom which is 2 in. wide to the top rear end which is 1⅜ in. wide. The bottom front end is smoothed off with a radius of ³¹⁄₃₂ in., and the rear end to a radius of ⅝ in. The thickness remains at 1⅜ in. throughout.

Fin Rib. There are four ribs in this fin. The top rib should be constructed of spruce cap strips ⁵⁄₁₆ in. thick and ⅜ in. wide, edge grain on the larger dimension. The top rib has an over-all length of 26¾ in. It is 2 in. wide at the rear and outside dimension, and is 1⅜ in. wide at the front and outside dimension. A jig will not be necessary to construct this rib. This rib should be covered on the top side with ply-

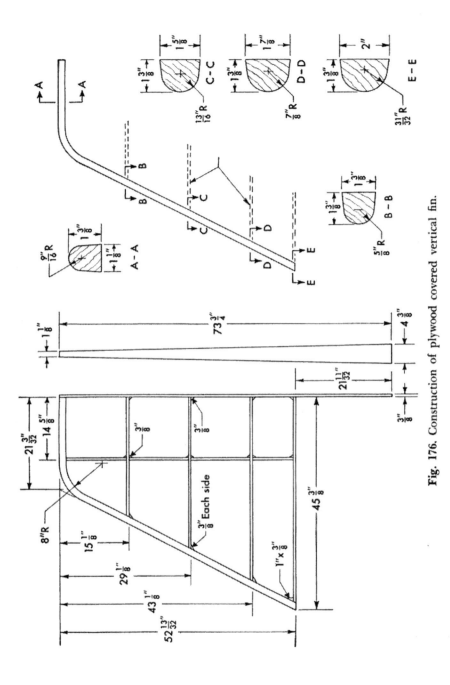

Fig. 176. Construction of plywood covered vertical fin.

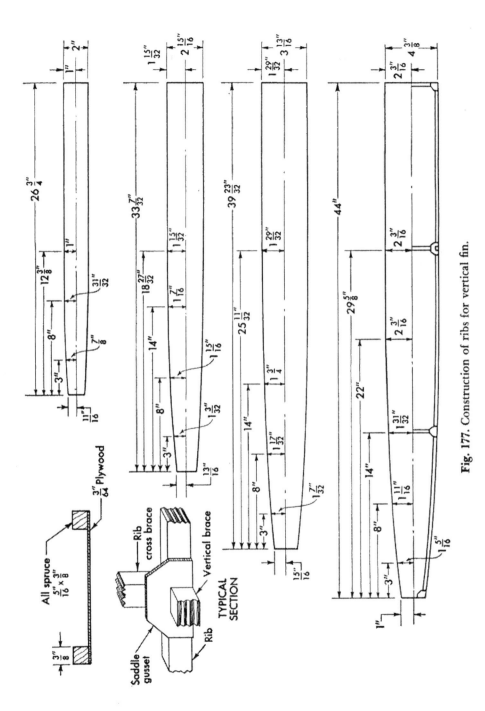

Fig. 177. Construction of ribs for vertical fin.

wood 3/64 in. thick. This plywood may be of any plywood available in this thickness, but mahogany outside layers are preferred. The plywood is cut the shape of the rib as shown in Figure 177. The end pieces of 3/8 in. by 5/16 in. spruce are cut to the proper length. A cross piece should be placed 12 3/8 in. from the small end of the rib as shown in Figure 177. The cap strip itself can now be bent into place and fastened at each end with two plywood gussets, as shown in Figure 177. The various cross pieces should be fastened in the same way with gussets of 3/64-in. plywood glued and nailed. The rib is then turned over, the top surface is coated with casein glue, and the plywood which has been cut to shape is nail-glued to the rib as shown in Figure 177. The other three ribs are constructed in a similar manner.

Vertical Bracing Members. There are two vertical bracing members which extend from the bottom rib to the top member of the fin. These strips are 3/8 in. by 5/16 in. spruce and are fastened in place by the use of saddle gussets and glue blocks, as shown in Figure 177.

Glue Blocks. The following glue blocks are required: 5 for fastening the ribs to the main spar at the end; 16 to fasten the vertical bracing members in place; and 4 to fasten the ribs to the leading edge. All of these blocks are prepared according to the dimensions given in Figure 176.

Assembly. The fin is assembled by nailing and gluing the four ribs to the main spar as shown in Figure 176. The top of the rear member is then glued and nailed to the laminated edge, followed by the front end of the ribs. With the assembly firmly clamped in place, insert and fasten the vertical braces as shown in Figure 176. Allow to dry not less than 8 hr.

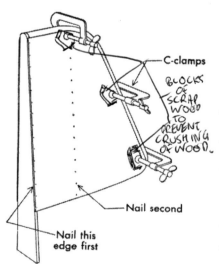

Fig. 178. Applying plywood covering to vertical fin.

Covering Fin. The whole fin is to be covered on both sides with plywood 3/64 in. thick, preferably mahogany outside veneer, although any plywood available of this thickness may be used. The plywood should extend from the rear edge of the main spar, and from the bottom edge of the bottom rib, and be of

PLYWOOD CONSTRUCTION

such size as to cover the leading and top edge up to the curved portion. The grain on the outside veneer should be placed vertically. The edge of the ribs, the main spar, and the leading and top edge should be coated with casein glue, and the plywood covering clamped along the leading and top edge with C-clamps, as shown in Figure 178. The plywood covering is then nailed in place, as shown in Figure 178, and allowed to dry not less than 24 hr. After thoroughly drying, the edge of the plywood along the front and top edge should be dressed down to make a smooth, tapered joint with the leading and top edge. This can be done with a fine wood rasp and sandpaper.

Construction of the Rear Section of a Plywood-Covered Fuselage. By using plywood as stressed-skin covering, it is possible to construct a fuselage entirely of wood. A fuselage having no diagonal cross members is called "monocoque construction." The construction of all bays of a fuselage of this type is similar throughout. The building of a part of a fuselage consisting of three rear bulkheads will familiarize the workman with this type of construction. The bulkheads should be laminated from $\frac{1}{32}$-in. mahogany-faced plywood, and each bulkhead should have five laminations, making a total thickness of $\frac{5}{32}$ in. The

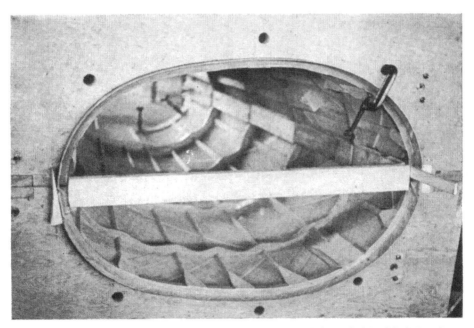

Fig. 179. Assembling a molded fuselage. (Courtesy Universal Moulded Products Corporation)

AIRCRAFT MAINTENANCE AND SERVICE

eight longerons are of clear hickory. Ash or Douglas fir may be substituted if hickory is not available. The covering should be of $\frac{3}{64}$-in. plywood, either birch- or mahogany-faced.

Longerons. The eight longerons are constructed of clear hickory, rectangular in shape, $\frac{5}{8}$ in. thick, $\frac{3}{4}$ in. wide; four being 36 in. long, and four being slightly longer due to the slope of the fuselage. These members should have edge grain on the wide face.

Bulkheads. It is not necessary to use a jig to construct the bulkheads. Each layer of the bulkhead may be built up of several pieces of plywood joined together with a scarf joint which should have a slope of 1 in 20. The joints should be so arranged that the joints in the various layers do not fall in the same places in the lamination. It would be possible to saw all laminations in a single piece, but this would cause a great waste of plywood. The sides of the plywood which are to be glued together should be lightly scraped or sandpapered to remove the glazed finish. The proper amount of casein glue should be spread on each face of the glued layers, and pressure applied to the entire bulk-

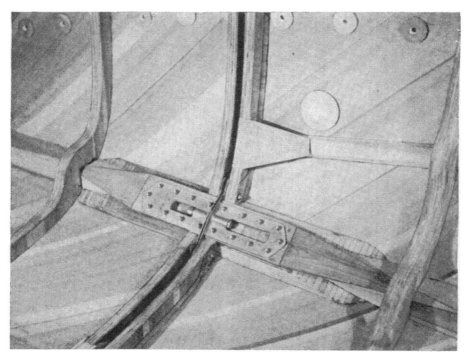

Fig. 180. The method of joining the molded parts of a fuselage. (Courtesy Universal Moulded Products Corporation)

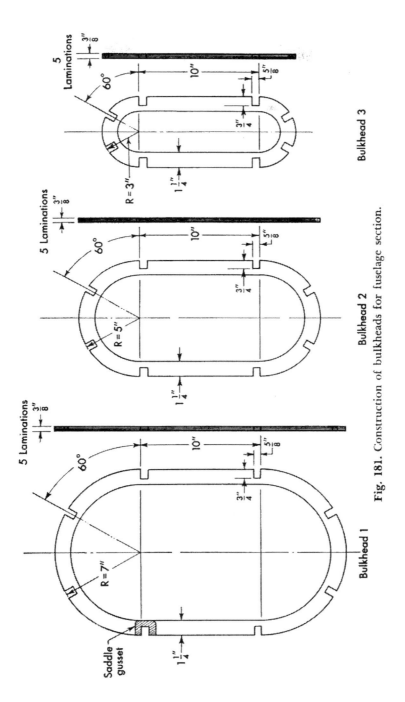

Fig. 181. Construction of bulkheads for fuselage section.

AIRCRAFT MAINTENANCE AND SERVICE

head. Approximately 175 lb. pressure per square inch should be applied. Allow 8 hr. for drying; 24 hr. additional drying should be allowed before notches for the longerons are cut. Reinforcing saddle gussets should be prepared as shown in Figure 175; 16 of these gussets will be necessary for each bulkhead.

Assembly. The structural part of the fuselage is assembled by gluing the longerons into the notches and clamping into place. The sad-

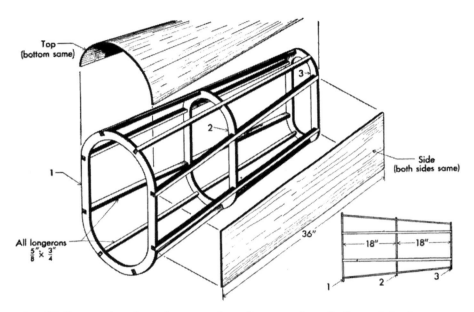

Fig. 182. Perspective view of construction of rear section of all wood fuselage.

dle gussets should be placed on each side of the bulkhead and nailed and glued as soon as the longerons are in place.

Covering. The plywood covering, which consists of $3/64$-in. mahogany-faced plywood, should be applied in four pieces. Two pieces, which cover the sides of the fuselage, should be flat and rectangular as shown in Figure 182; and the other two pieces should be preformed as shown in Figure 182. The sides of the pieces should be scarfed on the top and bottom edges and glued and nailed into place. The preformed plywood is then applied in a similar manner, forming a scarf joint with the side pieces. Care should be taken to obtain a good scarf joint. The plywood skin may be finished in an approved manner.

Figure 183 shows a form which may be built to shape the top and bottom pieces of plywood.

PLYWOOD CONSTRUCTION

Miscellaneous Construction. The construction used in these problems is typical of construction used in aircraft work.

Vertically-Laminated Spar. The spar is built up of five laminations; three vertical-grain and two flat-sawed pieces. The grain should be matched as shown in Figure 103 so that it extends in different directions in adjacent layers. The laminations will be ¼ in. thick and 6 in. wide. The fit between the pieces should be as nearly perfect as possible. Hardwood backing blocks should be prepared, each 1¼ in. in

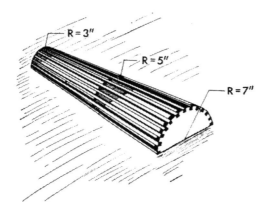

Fig. 183. Form for preforming plywood covering for top and bottom of fuselage section.

thickness and of sufficient size to cover the material being glued. A pressure of approximately 150 lb. per sq. in. should be applied.

Vertically-Laminated Spar Splice. A scarf joint should be prepared and glued. After gluing, the joints should be reinforced with plywood which is glued and nailed. All splices in spars should be reinforced. The plywood reinforcement on each side should be ¼ the thickness of the spar.

I-Beam Laminated Spar. This spar should be 8 in. deep, and the web formed from two laminations each ⁵⁄₁₆ in. thick. Each of the top and bottom flanges will be vertically laminated of four pieces ⅜ in. thick and 1 in. deep. This section of the spar is approximately 30 in. in length. The two pieces forming the web are prepared and glued together. After the web is thoroughly dry, the spar flanges are prepared and glued to the top and bottom of the web in a single operation.

I-Beam Spar Splice. A scarf joint is prepared and glued, and the proper reinforcing blocks are glued and nailed into place.

Plywood Box Spar. This spar is to be 8 in. deep, 4 in. wide, and approximately 30 in. long, built from $\frac{3}{32}$-in. plywood with corner blocks $1\frac{1}{8}$ in. deep and 1 in. wide, edge grain on the wide surface. The plywood for the two sides is cut to the proper shape, which will be 8 in. wide and 30 in. long. The surface of the plywood that is to be glued to the corner blocks should be lightly sanded to remove any gloss or glue on the surface of the plywood. The corner blocks are now prepared to exact size, and the plywood forming the sides of the spar is nail-glued to these blocks. There should be two rows of nails approximately $\frac{3}{4}$ in. apart, and the nails in the rows staggered. The nails should penetrate the block not less than 3 times the thickness of the plywood. The nails should be driven down firmly, but the heads must not be embedded in the wood. After allowing each side to dry not less than 8 hr., the top and bottom webs are nail-glued into place. The webs should be allowed to dry not less than 8 hours.

Cap-Strip Bending Form. In making repairs to the nose portion of the damaged ribs, it is not always desirable to build a complete jig when replacing a small portion of the cap strips of the rib. It is difficult to replace this portion of the rib without a jig. The top and bottom cap strips should be bent before making the repair. Figure 219 shows the method of construction of this bending form, using a shaped block and hand clamp.

Cap Strips. Using the cap-strip bending form, prepare three cap strips for repairs to the nose section of a rib. Prepare cap strips of the same material and the same size as the cap strips used in the ribs. Soak one end of these strips in hot water until flexible, 30 min. should be sufficient, and place side by side in the cap-strip bending form. Tighten on the form by means of two C-clamps and a cross piece. Just enough pressure should be exerted to get a snug fit on the form. Excessive pressure may deform the strips. Allow to dry approximately 24 hr. Use these cap strips in making repairs to the ribs.

Inspection Plate. Install an inspection plate in the plywood covering of the wing tip which has already been constructed. The method for installing this plate is the same as for installing a plug patch except that the cover is not glued into place. The plate is fastened into place with screws provided with plate nuts.

An inspection plate may be installed as follows. Prepare a frame of spruce or other suitable material. Cut the proper-sized hole in the fabric, turning back the fabric over the spar and rib. Glue and nail the

PLYWOOD CONSTRUCTION

inspection frame into place. Prepare an inspection plate of $3/64$-in. plywood and fasten into place on this frame with screws having plate nuts which have been installed on the frame. Bring the fabric to the edge of the inspection opening and stitch to the inspection hole frame. Dope into place and reinforce with tape.

XVII TYPICAL WOOD AND FABRIC REPAIRS

Straight Tear in Fabric. To repair a straight tear in a fabric covering, remove the finish from the fabric with a dope thinner or some suitable solvent for at least 1 in. in all directions from the damaged portion. The edges should be drawn together using a baseball stitch. The thread should be the same as the thread used for sewing the fabric covering and should be locked by using the seine knot. The patch should extend 1 in. beyond the edge of the tear in all directions, and the edges may be pinked, frayed, or plain. The patch should be of the same material as the fabric being repaired. A coat of clear dope should be applied over the repair, and the patch pressed into the dope while it is still wet. The patch is then covered with a coat of clear dope and allowed to dry 30 to 45 min. The repaired area may then be finished in the same way as the original fabric.

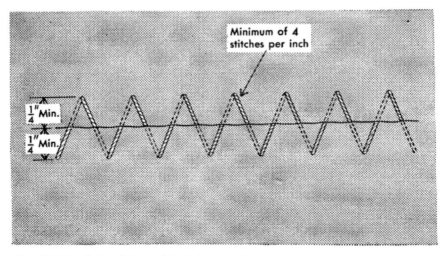

Fig. 184. Baseball stitch used in fabric repair.

TYPICAL WOOD AND FABRIC REPAIRS

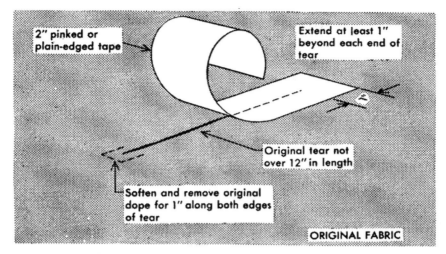

Fig. 185. Patch for small tears not over 12 inches long.

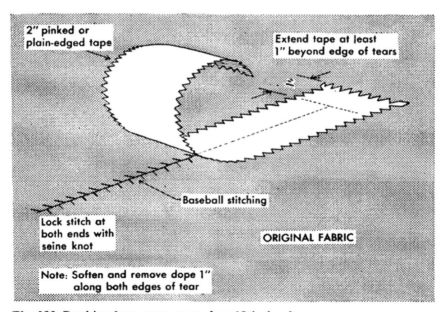

Fig. 186. Patching large tears more than 12 inches long.

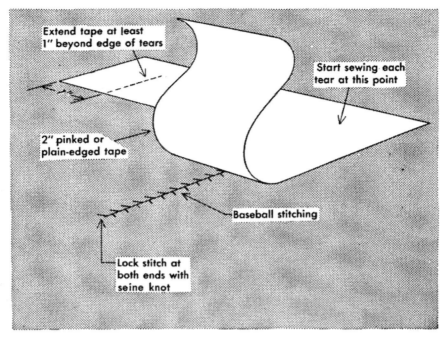

Fig. 187. Patching large "V" tears more than 6 inches long.

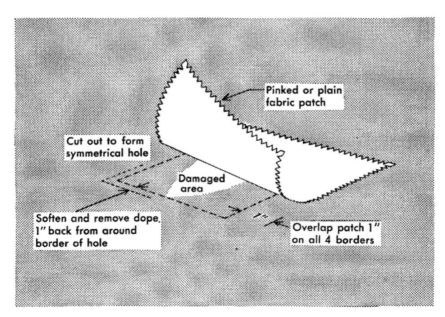

Fig. 188. Patching small holes in aircraft fabric covering.

TYPICAL WOOD AND FABRIC REPAIRS

Triangular Tear in Fabric. To repair a triangular tear, remove the original finish from the area to be covered by the patch with dope thinner or other solvent. The edges of the tear are then drawn together using a baseball stitch as shown in Figure 184. The patch should then be prepared similarly to the straight-tear patch and should extend beyond the repair at least 1 in. on all sides.

Hole in Fabric. Small holes in fabric, having a diameter of less than 3 in. in any direction, may be repaired in the following manner. Cut out any ragged edges. Remove the original finish for at least 1 in. in all directions from the edge of the hole. A patch extending at least 1 in. in all directions from the hole is applied in the manner described in the first paragraph of this chapter.

Wing-Tip Damage to Fabric. The wing tip is very often damaged. This damage may be repaired by removing the tape from the end of the wing for a distance of not less than 1 in. beyond the damaged area. Remove the finish for at least 1 in. around the damaged area by use of dope thinner or other solvent. Stitch the fabric and recover with tape, doping in place as directed above.

Trailing-Edge Damage to Fabric. Remove the reinforcing tape covering the trailing edge. Remove the original finish from the fabric, and draw the fabric into place by stitching. Dope the reinforcing tape over the repaired area.

Large Area of Damaged Fabric. Remove the damaged fabric by cutting out the fabric to within 2 in. of the ribs adjacent to the part to be re-covered. Remove the original finish from the fabric for a distance of not less than 2 in. beyond the rib adjacent to the repair. In repairs of this type, the new fabric should extend not less than 3 in. beyond the leading edge, 2 in. beyond the ribs adjacent to the repaired area, and approximately ½ in. around the trailing edge. The fabric is then stretched over the area to be repaired, as shown in Figure 189, and the edges doped to the fabric already in place. After the edges have dried, the new piece of fabric is rib-stitched to the ribs beginning with those adjacent to the repaired area, each rib being covered with a reinforcing tape in the same manner as applying new fabric. The edges of the newly applied fabric should be covered with pinked tape, and the whole doped and finished in an approved manner.

Emergency Repair of a Small Hole in Plywood. An emergency patch may be applied to a small hole in a plywood covering in the following manner. With a sharp knife or other suitable tool, trim out the

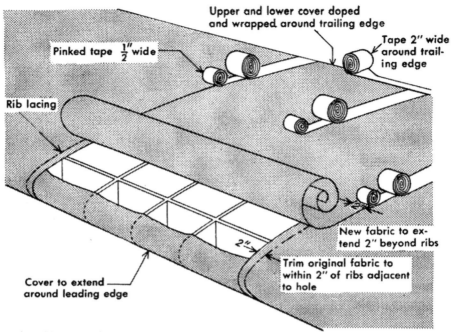

Fig. 189. Replacing large areas of fabric.

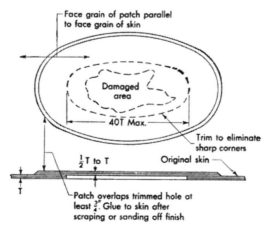

Fig. 190. Temporary patch to plywood covering.

TYPICAL WOOD AND FABRIC REPAIRS

damaged wood to eliminate any sharp corners. Scrape the finish off the plywood to a distance of approximately 1 in. from all sides of the hole. Prepare a plywood patch of such size that it will extend not less than 3/4 in. on all sides of the hole. Bevel the edge of the patch. The grain of the patch should be parallel to the grain of the plywood being repaired. The patch may now be glued in place, pressure being applied by a block, if the surface is level, or by means of a sand bag on curved surfaces. A piece of paper should be placed over the patch to prevent the sticking of the glue to the block or sack used to apply pressure. After gluing, the paper may be removed by sanding.

Emergency Repair of a Large Hole in Plywood. Emergency repairs to large holes in a plywood covering may be made as shown in Figure 191. When the leading edge is damaged, the damaged plywood should be cut out until the edge of the patch will rest on two adjacent ribs. The patch should extend around the leading edge, both top and

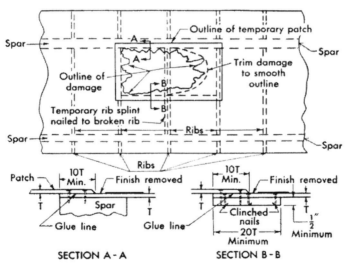

Fig. 191. Temporary patch to plywood wing surface.

bottom, for a distance of not less than 6 or 8 in. The patch should then be nailed and glued into place, soaking the patch if necessary. Use the leading edge for a form to shape the patch if the plywood has been soaked. Allow to dry before gluing and nailing into place. It is often advisable to dope a piece of fabric over a temporary patch of this kind. The doped patch should dry not less than one hour. Figures 192 and 193 show the steps in making this type of repair.

AIRCRAFT MAINTENANCE AND SERVICE

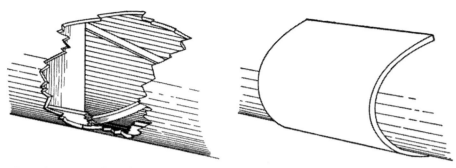

Fig. 192. Damaged leading edge.

Fig. 193. Temporary patch on leading edge.

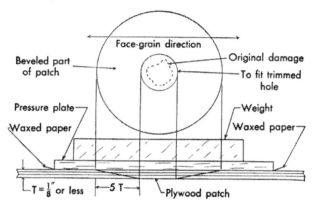

Fig. 194. Splayed patch.

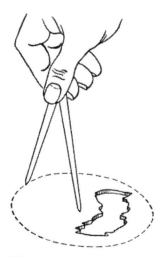

Fig. 195. Cut out damaged area in form of circle.

TYPICAL WOOD AND FABRIC REPAIRS

Permanent Repair to Small Holes in Plywood. Small holes may be repaired permanently by either a splayed patch or a plug patch. A splayed patch is a patch with tapered edges, but the slope is steeper than a scarfed edge. In a splayed patch the taper is approximately 5 times the thickness of the material. Holes repaired by this method should not exceed in diameter 20 times the thickness of the skin. The hole is cut out in the form of a circle, as shown in Figure 195. The edge

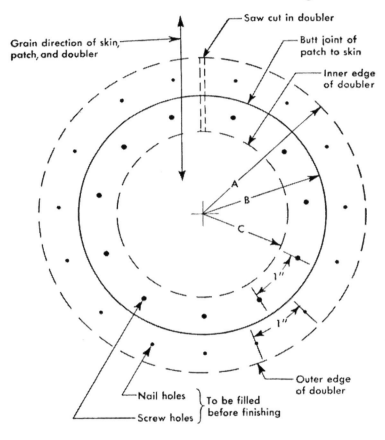

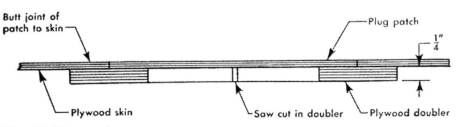

Fig. 196. Plug patch.

249

of the hole is then tapered evenly for a splayed patch, using either a sharp knife, chisel, or a wood rasp. If a plug patch is used, the edge of the hole is trimmed at right angles to the surface of the plywood, as

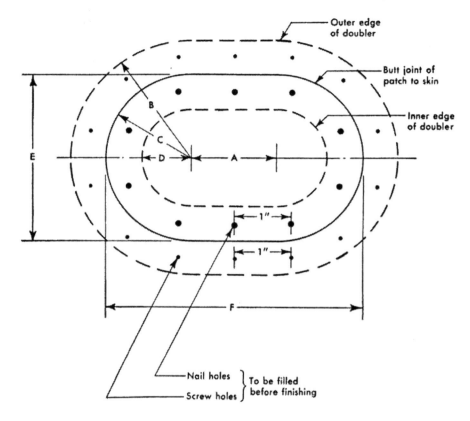

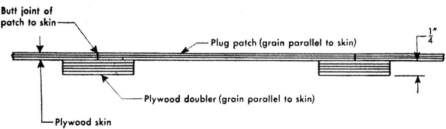

Fig. 197. Oval patch.

shown in Figure 196. The edge of the patch is then trimmed to fit the hole exactly and glued in place. The entire patch should be covered with waxed paper or some suitable material to keep the glue from sticking to the clamps or weights used to apply pressure.

TYPICAL WOOD AND FABRIC REPAIRS

An oval plug patch may be used for holes up to approximately 5 to 7 in. in size. The holes should be trimmed to the correct oval shape, and the finish removed from both the inside and outside of the plywood

Fig. 198. Trimming a hole for an oval patch.

Fig. 199. Removing finish from plywood.

covering for at least 1 in. from the edge of the hole, as shown in Figures 197 and 198. In applying the oval patch, a doubler is used. A doubler is a reinforcement placed under the edge of the hole inside the skin. This doubler is nailed and glued in place using nailing strips.

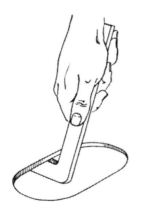

Fig. 200. Removing finish from the underside of plywood with scraper.

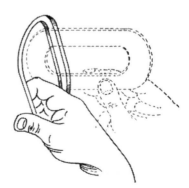

Fig. 201. Applying doubler for oval patch.

The patch is cut to the exact size of the hole and glued and screwed to the doubler, using a pressure plate. On larger sized patches, two rows of nails or screws should be used. After the patch has thoroughly dried, remove the nailing strips with a pair of pincers. Scrape off the

AIRCRAFT MAINTENANCE AND SERVICE

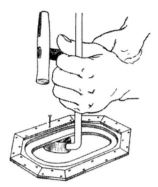

Fig. 202. Nailing strips to hold doubler in place.

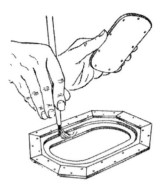

Fig. 203. Applying glue for oval patch.

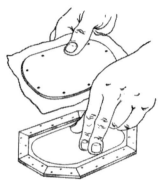

Fig. 204. Oval patch in place; pressure plate in background.

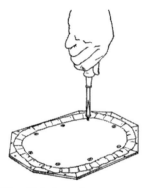

Fig. 205. Screwing pressure plate into place.

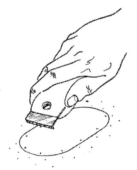

Fig. 206. Removing excess glue with scraper after pressure plate and mailing strips have been removed.

Fig. 207. Screw holes filled with plastic wood and sandpapered smooth.

TYPICAL WOOD AND FABRIC REPAIRS

excess glue and finish smooth with sandpaper. Figures 208 and 209 show the placing of a doubler for a round plug patch by first splitting the doubler on one side.

Plywood Scarf Patches. Damaged areas too large to be covered by oval or round patches should be repaired by inserting a scarf patch.

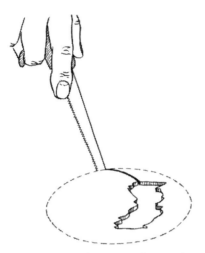 

Fig. 208. Sawing out damaged area in form of circle.

Fig. 209. Split doubler being put in place.

Scarf patches may be used in any place in the plywood skin except that no horizontal scarf joint is allowed on the leading edge. In sharply curved parts it is necessary to preform the plywood to be used before scarfing its edges. Patches covering an area extending over two or more ribs should have reinforcing strips placed along the edge of the ribs, as shown in Figure 212. The edge of the plywood may be scarfed by a hand plane, scraper, accurate sanding block, or chisels. The slope used on a scarf patch should not be greater than 1 in 15. The corner of a hole for a scarf patch should be rounded, and proper backing strips should be placed under the edges of the plywood skin. Nailing strips should be used to fasten the backing strips into place. Backing strips should be made of a softwood such as spruce or poplar. The edges of the scarfed end of the patch are glued and nailed into place using hot sandbags to apply pressure. Plywood for a patch on the leading edge should be preformed either by using the leading edge of the wing as a form, or as shown in Figure 218. Place heavy burlap or paper under the plywood to prevent injury to the leading edge. Backing strips for the

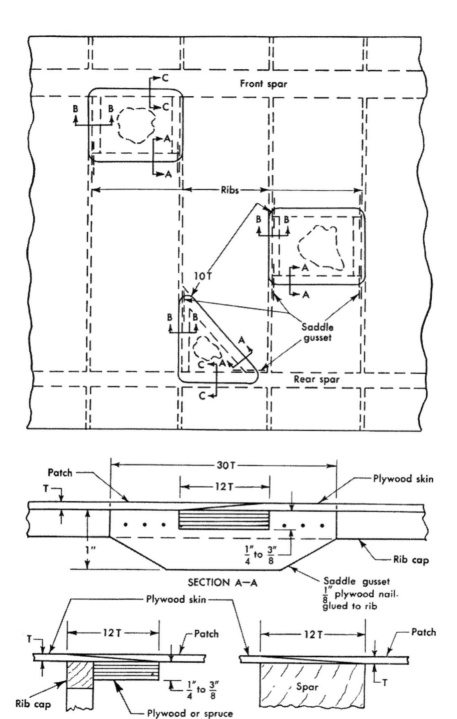

Fig. 210. Scarf patches on wing surface.

TYPICAL WOOD AND FABRIC REPAIRS

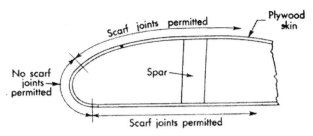

Fig. 211. No scarf joints allowed on leading edge of wing.

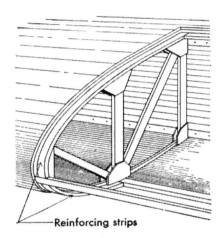

Fig. 212. Reinforcing strips along edge of ribs.

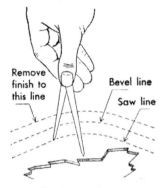

Fig. 213. Rounding corners for scarf patch.

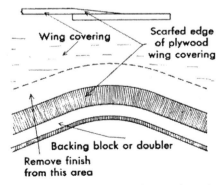

Fig. 214. Backing strips for scarf patch.

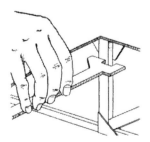

Fig. 215. Backing strips to hold scarf patch in place.

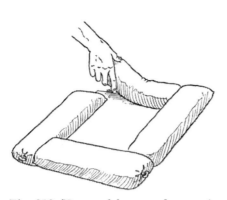

Fig. 216. Hot sand bags used to apply pressure.

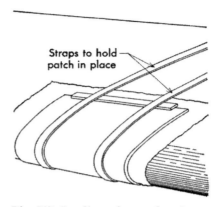

Fig. 217. Leading edge used to form patch.

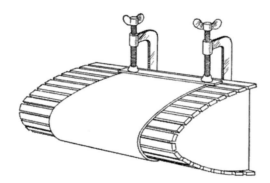

Fig. 218. Forming leading edge patch.

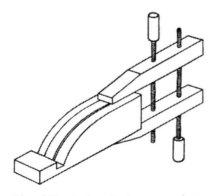

Fig. 219. A hand clamp used to form a backing strip.

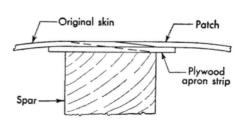

Fig. 220. Scarf joint over an apron strip.

TYPICAL WOOD AND FABRIC REPAIRS

leading edge may be formed by using hand clamps as shown in Figure 219. After the patch has thoroughly dried, nailing strips should be removed with a pair of pincers and the surface properly finished by scraping off excess glue and finishing with sandpaper. Figure 220 shows the repair of a scarf joint over the apron strip.

Wing-Structure Approved Repairs. The wing framework consists of spars, ribs, leading-edge strip, wing-tip bow, trailing-edge strip, compression members, braces, and reinforcing members. A scarf joint is the most satisfactory method of making a repair to a solid wood member. The joint should be cut accurately and the same slope be made on each of the pieces to be joined.

Spar Repairs. A spar may be spliced at any point except under wing fittings or where other parts, such as landing gear fittings and engine fittings, are attached to the spar. Reinforcing plates must be used on all scarf repairs to spars. The proper slope of a scarf joint for a solid or

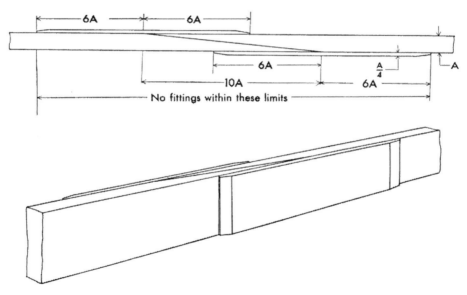

Fig. 221. A method of splicing solid rectangular spars.

rectangular spar should not be less than 10 times the thickness of the part spliced. Figure 221 shows how this splice should be made. Either end of the spar may be replaced. The fabric covering of the wing should be loosened along the trailing edge and completely removed from the wing. The wing walk should be removed. The plywood leading edge should be loosened from the front spar. All fittings should be loosened and removed from the front spar. The blocks on top of the front spar are next removed by the use of a chisel and hammer. The ribs are loosened from the front spar by removing the gussets at the top and bottom of the vertical-rib members. Using either a knife blade or a putty knife, the root of the front spar is loosened from the root rib and may be removed by sliding outward. After the splice has been made in the spar, as shown in Figure 221, the spar should be replaced and the wing rebuilt as before. The vertical members of the ribs are replaced, making close contact with the reinforcing blocks, and new gussets glued and nailed into place.

Cap-Strip and Rib Repair. A cap strip of a wood rib may be repaired by means of a scarf joint which is reinforced on the side opposite the wing covering by a spruce block which extends beyond the scarf joint not less than 3 times the thickness of the strips being repaired. The entire joint, including the reinforcing block, is reinforced on each side by a plywood reinforcing as shown in Figure 222. When the cap strip is to be repaired at a point where there is a joint between it and cross members of the rib, the repair is made by reinforcing the scarf joint with plywood gussets, as shown in Figure 223. When it is necessary to make a repair to a cap strip at a spar, the joint should be reinforced by means of a continuous gusset extending over the spar as shown in Figure 224. Edge damage, cracks, or other local damage in a spar may be repaired by removing the damaged portion and gluing in a properly fitted block, as shown in Figure 225, reinforcing the joint by means of plywood or spruce blocks which are glued into place. The trailing edge of a rib may be replaced and repaired by removing the damaged portion of the cap strip and inserting a softwood block of white pine, spruce, or basswood. The entire repair is then reinforced with plywood gussets and nailed and glued as shown in Figure 226. A compression-rib repair may be made as shown in Figure 227. Compression ribs are of many different types, and the mechanic must judge the proper method for making a repair to any part of this type of rib. Figure 227 shows a repair made to a compression rib built up of a plywood web and three longitudinal members, the center one of which has been repaired by a properly reinforced scarf joint and an outside layer of plywood.

Glue Blocks, Filler Blocks, and Compression Members. Such members as glue blocks, filler blocks, compression members, braces, and rib diagonals should not be repaired, but should be replaced. Wherever

TYPICAL WOOD AND FABRIC REPAIRS

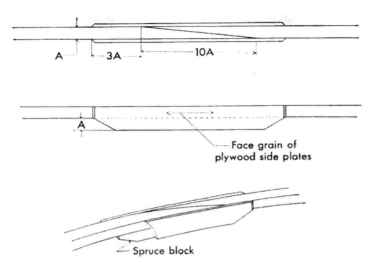

Fig. 222. A rib cap-strip repair.

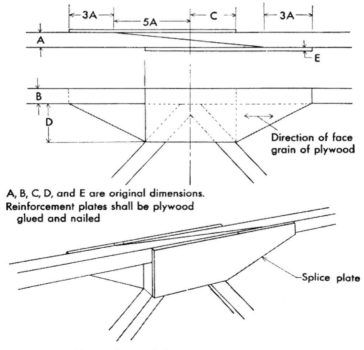

A, B, C, D, and E are original dimensions.
Reinforcement plates shall be plywood
glued and nailed

Fig. 223. A rib repair at a joint.

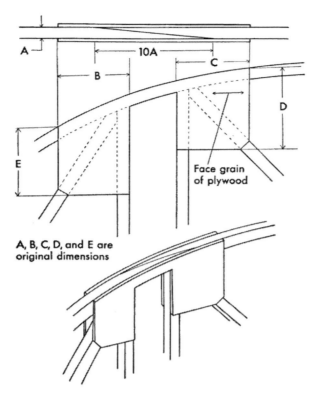

Fig. 224. A rib repair at a spar.

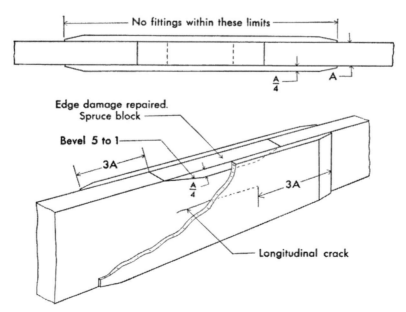

Fig. 225. The repair of cracks and edge damage on solid spar.

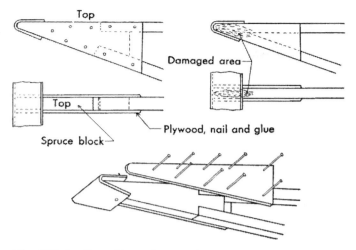

Fig. 226. A rib trailing edge repair.

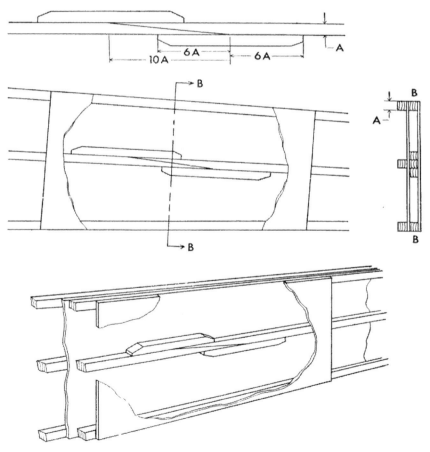

Fig. 227. A compressing rib repair.

it is possible to replace a damaged member, it is always better to do so than to attempt a repair.

Wing-Tip Bows. Wing-tip bows and other laminated structures, when it is impossible to replace the entire member, may be repaired by sawing out the damaged portion and splicing in by means of a scarf joint a preformed laminated section. A temporary repair may be made by splicing in a solid piece with a scarf joint, if the damage is not too extensive.

Finishing Repairs. After repairs have been made, the proper finish should be applied. When making repairs, care should be taken that the surrounding areas or the area repaired is not covered with glue or any other substance which will interfere with proper finishing.

Oil or grease may be removed with naphtha. Marks made by lumber crayons or grease pencils should be removed. Any chips, shavings, sawdust, glue particles, or other waste material should be removed from enclosed spaces. A vacuum cleaner may be used to remove such material.

End-grain surfaces and plywood edges should be given two coats of sealer. Drilled holes should be finished with two coats of sealer and one coat of aluminized varnish. Nail heads, screw heads, etc., should be covered with a strip of tape or other material to prevent cracking of the finish. Holes that are left by removing nails when nail-strip gluing has been used should be filled with plastic wood.

All new woodwork should be given not less than two coats of sealer. Exposed surfaces should be refinished to match the original finish.

Control Surface Repairs. Control surface repairs are made in the same manner as wing repairs.

Approved Fuselage Repairs. Fuselage repairs include repairs to longerons, stringers, cross braces, and vertical braces. These members are repaired by scarfing and reinforcing. Care should be taken that the grains are properly matched in making these repairs.

Bulkhead Repair. In repairing a bulkhead, a rectangular piece of plywood skin large enough to allow the repair to be made should be removed, as shown in Figure 228. The bulkhead is then scarfed, and a repair piece prepared of the proper shape. This piece is glued into place, and reinforcing members are applied. A doubler is placed around the edge of the hole cut in the plywood and a scarf patch applied. Figures 231 and 232 show the use of an adjustable curve for reproducing the curved portions of bulkheads or skin stiffeners.

TYPICAL WOOD AND FABRIC REPAIRS

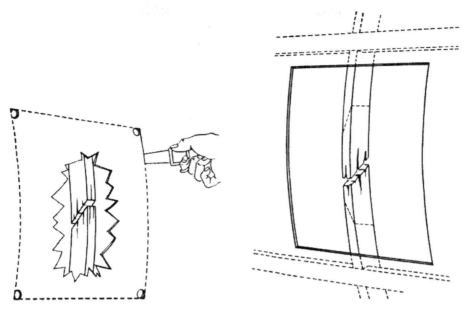

Fig. 228. Removing damaged material from plywood fuselage.

Fig. 229. Scarf ends of structural member.

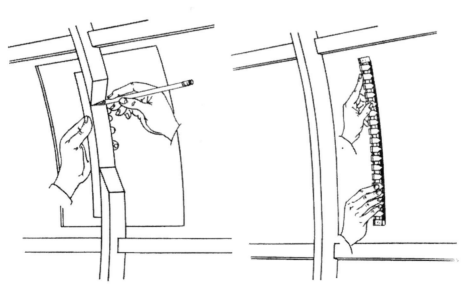

Fig. 230. Marking a piece of material to be used in repair of bulkhead.

Fig. 231. Using an adjustable curve to get the curve of the fuselage covering.

AIRCRAFT MAINTENANCE AND SERVICE

Repair of a Skin Stiffener. Skin stiffeners are repaired either by scarf splicing in a part to replace the damaged material, or by removing that portion of the skin stiffener between bulkheads and replacing as shown in Figure 233. The ends are strengthened by the use of saddle gussets.

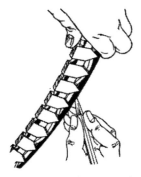

Fig. 232. Using an adjustable curve to mark material for the repair of a bulkhead.

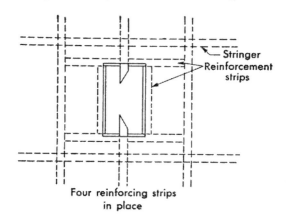

Fig. 233. A repair to a skin stiffener.

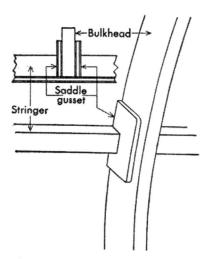

Fig. 234. Saddle gusset to reinforce a skin stiffener.

Bearing Blocks and Plywood Reinforcements. Bearing blocks and plywood reinforcements are repaired or replaced in the same manner as described for wing-structure repairs.

Fairings. Fairings do not carry structural loads. Damaged material may be replaced by gluing in filler blocks of the same material, or the entire fairing may be replaced.

XVIII WELDED REPAIRS

Joints Used in Aircraft Welding. The joints used in aircraft welding vary from a simple butt weld of two pieces of sheet metal to the complicated cluster welds of aircraft tubing. The thickness of the reinforcement varies with the thickness of the sheet being welded. On very thin stock the reinforcement may be as much as the thickness of the sheet, while on stock 1/8 in. thick the reinforcement may be 30 per cent of the thickness, and on material 1/4 in. thick the reinforcement may be reduced to approximately 20 per cent of the thickness of the sheet.

Much of the framework on light airplanes consists of welded tubular structures. Many fittings are welded into place on these structures. Engine mounts, vertical fins, rudders, stabilizers, landing gear, and the main fuselage framework in most light airplanes are of welded construction. It is necessary to use certain specified types of welds throughout aircraft construction.

Figure 238 shows a square-edged butt weld for sheet or tubing up to 0.093 in. in thickness. The penetration of the weld should be complete. The reinforcement should be approximately six times the thickness of the material in width and approximately the thickness of the material in depth. The space allowed between the edges to be welded should be approximately equal to the thickness of the sheet. Care should be taken, particularly with thin material, to avoid burning or melting through. Cold laps, rolled edges, and lack of proper reinforcement should be avoided. On thin material there is a tendency on the part of the inexperienced welding operator to form welds containing these defects, in order to avoid melting through or burning the material.

Figure 239 illustrates the fillet-weld angle joint to be used with sheet metal. This type of weld is not entirely satisfactory when the material welded is subject to corrosion, or when flux is used in making the weld. The flux flows into the joint and cannot be removed after the weld is completed. The weld should be built up on each side of the sheet to

Fig. 235. Welding an aircraft rudder framework. (Courtesy Piper Aircraft Corporation)

Fig. 236. Welding a steel fuselage. (Courtesy The Linde Air Products Company)

Fig. 237. Factory welding of light airplane fuselages. (Courtesy The Linde Air Products Company)

AIRCRAFT MAINTENANCE AND SERVICE

approximately 1½ times the thickness of the sheet. The base of the weld should be built out to approximately the same distance. Good penetration should be obtained on both the base sheet and the angle sheet, but the weld should not penetrate completely through the sheet forming the angle. A penetration of approximately ¼ the thickness of the sheet is sufficient. The edge of the weld should taper into the mate-

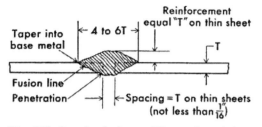

Fig. 238. Standards for welding a butt joint in thin sheet.

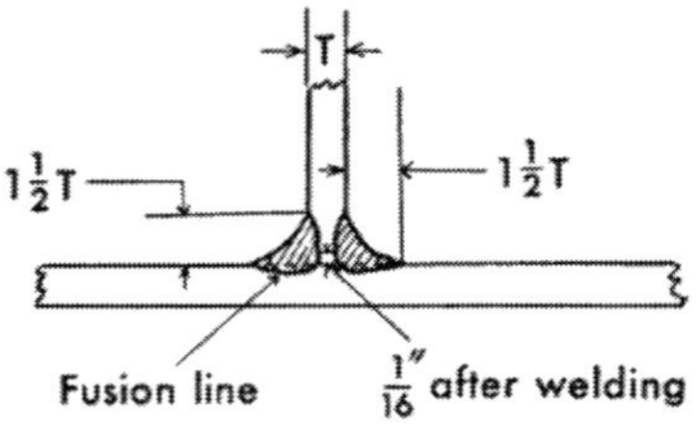

Fig. 239. Standards for welding "T" joint in thin sheet.

rial and be free from defects. Care should be taken that undercutting does not occur along the top of the fillet weld, or a rolled edge at the bottom of the weld. On all fillet welds, care should be taken that the material in both pieces is thoroughly fused down to the root of the joint. Fillet welds usually have a slightly concave surface.

Figure 240 shows a fillet lap joint which may be used for either sheet or tubing. This joint is not as desirable as the plain butt joint, particu-

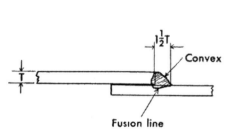
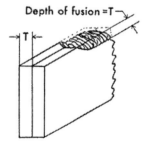

Fig. 240. Standard lap weld on thin sheet.

Fig. 241. Edge weld to be used on fittings formed from sheet.

larly for metals which are subject to corrosion. When flux is used on this type of joint, it may penetrate the joint and be unable to be removed. Joints of this type are not superior in strength to a plain butt weld. The surface of this fillet weld should be convex. The penetra-

WELDED REPAIRS

tion should be carried down into the joint between the metals and both pieces thoroughly fused.

Figure 241 shows an edge weld to be used on sheet or fittings formed from sheet. This weld may be made either with or without the addition of metal. If no metal is added, the edge of the sheet should be melted down to such a depth that the fused metal is equal in depth to the thickness of the sheets being welded.

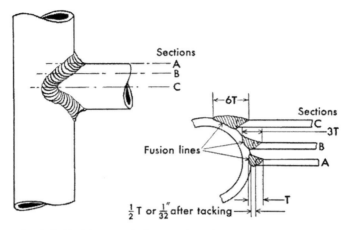

Fig. 242. Tubing welded to tubing showing proper penetration at various points in the weld.

Figures 242 and 243 show the method of welding either sheets or tubes to other tubes. Care should be taken that the penetration is as great as shown in the figure. When the fillet is made from only one side, the penetration must be completely through the material welded to the tube.

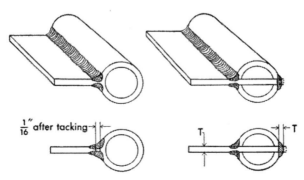

Fig. 243. Welding sheets to tubes.

269

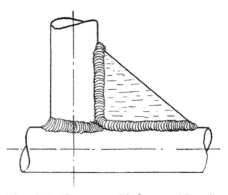

Fig. 244. Gusset welded to tubing by scale welding.

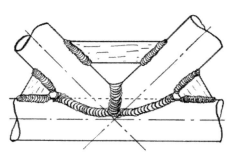

Fig. 245. Tube cluster reinforced with gussets.

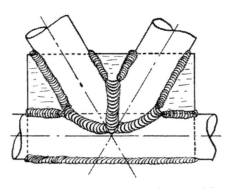

Fig. 246. Tube cluster reinforced with inserted and welded gusset.

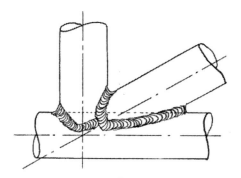

Fig. 247. Three tube cluster.

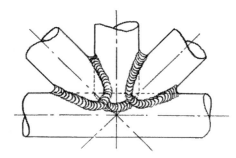

Fig. 248. Four tube cluster.

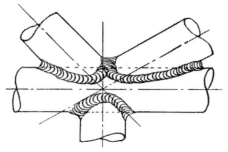

Fig. 249. Four tube cluster.

WELDED REPAIRS

Figure 244 shows the method of welding gussets to tubing.

Figures 245, 246, 247, 248, 249, and 250 show the various methods of forming joints between tubing and the reinforcement of these joints by means of gussets. It should be noted that the center line of each tube meets at a common point with the center line of the tube to which it is welded. This arrangement prevents the load on the various members from developing tear stresses. It also prevents torsional or bending stresses upon the main member. Highly stressed joints are usually reinforced with gussets which may be fitted into slots sawed into the

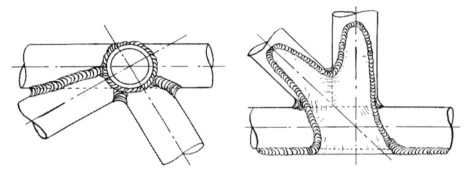

Fig. 250. Four tube cluster with one tube at right angles to the others.

Fig. 251. Reinforcing finger patch.

tubes or welded externally to them. The gussets may be bent around the main member and welded to the tubes joining the main member, as shown in Figure 251.

Figure 252 illustrates the various types of welded seams used in the construction of fuel and oil tanks. These seams may be in the form of butt joints, corner joints, edge joints, or lap joints. When baffle plates are riveted to the outer shell of a tank, the rivet heads may be welded to form a liquid-tight joint. In welding seams in tanks, a corrugation is often formed along the seam to take care of contraction and expansion. These corrugations will also add stiffness to the shell.

Joint A in Figure 252 illustrates the welding of rivet heads.

Joint B is a method sometimes used to insert baffle plates. In this joint the edges of the shell are bent at right angles to the shell, and the baffle plate welded into the slot thus formed.

Joint C shows a locked seam which has been sealed by welding.

Joint D is another style of seam which may be welded either with or without the addition of metal to the joint.

AIRCRAFT MAINTENANCE AND SERVICE

Joint E is an edge joint which may be formed either with or without the addition of metal to the weld.

Joint F illustrates a type of corner seam which may be formed either with or without the addition of metal to the weld.

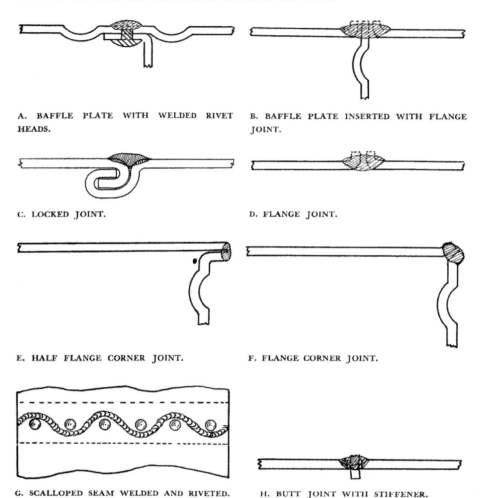

A. BAFFLE PLATE WITH WELDED RIVET HEADS.

B. BAFFLE PLATE INSERTED WITH FLANGE JOINT.

C. LOCKED JOINT.

D. FLANGE JOINT.

E. HALF FLANGE CORNER JOINT.

F. FLANGE CORNER JOINT.

G. SCALLOPED SEAM WELDED AND RIVETED.

H. BUTT JOINT WITH STIFFENER.

Fig. 252. Various types of welded seams and joints used in tank construction.

Joint G illustrates a scalloped seam which is riveted and welded. In this joint the rivets must also be sealed by the weld.

Joint H illustrates a butt joint in which a stiffening strip has been placed to add strength to the weld.

Figure 253 shows a butt weld for aircraft tubing. The space left between the parts should be approximately equal to the thickness of the

WELDED REPAIRS

wall of the tube being welded. The joint should be tack-welded in three places equally spaced around the tube. The weld is then carried around the tube, fusing out the tack welds. The cross section of the weld should be the same as for a butt weld of a plate.

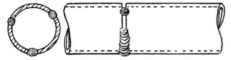 

Fig. 253. Butt joint in tubing. Fig. 254. Scarf joint in tubing.

Figure 254 illustrates a scarf weld. The slope of the weld should be not less than the diameter of the tube and may be as great as twice the diameter. Space should be left between the parts equal approximately to the thickness of the tube wall. It is good practice to tack-weld this type of joint in four places equally spaced around the weld. The weld may then be made continuously around the tube, melting out the tack welds.

Figure 255 shows a fishmouth joint. The angle formed by the joint should be not less than 60 degrees. This type of joint should be tack-welded at about four points spaced an equal distance apart. The weld

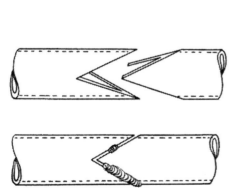 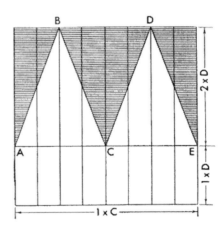

Fig. 255. Fishmouth joint in tubing. Fig. 256. Template for laying out fishmouth joints.

may then be carried continuously around the tube. A cross section of the weld and reinforcing should be the same as that described for a butt weld in sheet. Figure 256 shows the laying out of a template used in

273

preparing a fishmouth joint. Figure 257 shows a fishmouth weld which has been reinforced with an internal sleeve.

Rosette welds should be made as shown in Figure 257. Rosette welds are formed by drilling a hole through the outside tube before inserting the inner sleeve. These holes are then welded full. The inner tube should be fused, and the penetration should extend beyond the joint

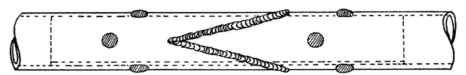

Fig. 257. Fishmouth joint in tubing with inner sleeve and rosette welds.

formed by the bottom of the drilled hole. Enough metal should be added to slightly reinforce the weld. Rosette welds fasten the inner tube similarly to the main member. The rosette weld should be made after the main fishmouth weld has been completed. This relieves

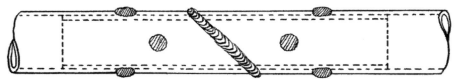

Fig. 258. Scarf joint in tubing with inner sleeve and rosette welds.

stresses which might be placed on the rosette weld due to the expansion of the main member during the welding of the fishmouth joint.

Figure 258 shows a scarf splice and rosette welds with an inner-tube reinforcement.

Figure 259 shows a double fishmouth weld on a reinforcing sleeve.

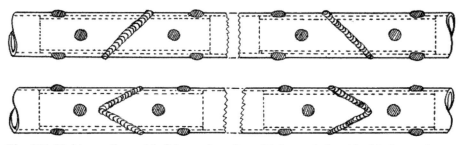

Fig. 259. Tubing splices with fishmouth and scarf joints reinforced with inner sleeve and rosette welds.

WELDED REPAIRS

Figure 260 shows a reinforcing sleeve which has been split lengthwise and welded into place.

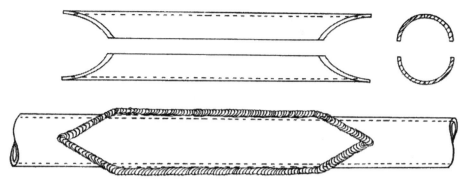

Fig. 260. Reinforcing sleeve split and welded around tube.

Figure 261 shows a reinforcing plate used for strengthening a junction of truss members. The sheet-metal plate of the same material as the truss members is first cut to shape and tack-welded into place. It is then formed around the tubes by pounding into place and welded, using a fillet weld.

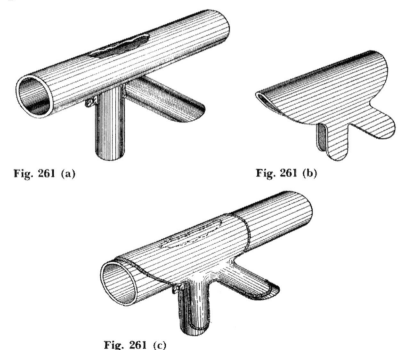

Fig. 261 (a)   Fig. 261 (b)

Fig. 261 (c)

Figs. 261 (a, b, c). Formed patch for strengthening damaged tube cluster.

AIRCRAFT MAINTENANCE AND SERVICE

Figure 262 shows a gusset plate inserted through a tube to reinforce a scarf splice. The scarf splice is first formed, then the ends of the tube are slotted and the gusset inserted. The scarf splice between the sections of the tube should be welded first. The gusset plate is then welded

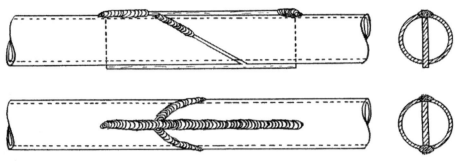

Fig. 262. Gusset inserted through tube to reinforce scarf splice.

along each edge. Metal should be added by means of the proper-sized filler rod.

Cleaning and Finishing Welded Joints. A great many broken parts to be welded have been covered by some kind of coating to prevent rust and other forms of corrosion. These coatings may be paint, varnish, enamel, or various types of metal plating and must be removed before welding, brazing, or soldering. Dirt, grease, and oil must also be thoroughly removed.

Cadmium plating may be removed by dipping the edges to be welded in one of the following solutions:

1. 73 cu. cm. of hydrochloric acid, 27 cu. cm. of water, and 2 g. of antimony trioxide.
2. 1 lb. of ammonium nitrate and 1 gal. of water.
3. 3 gal. of water, 7½ gal. of hydrochloric acid, and 1½ lb. of ammonium nitrate.

Paint, varnish, and enamel may be removed by buffing, sandblasting, or a paint and varnish remover. A 10 per cent solution of hot caustic soda or tri-sodium phosphate will remove most paints and varnishes. Any corrosion present on steel parts should be removed by sandblasting or buffing with a wire brush. It is always advisable where strong caustic soda or tri-sodium phosphate has been used to treat the part after cleaning with a 10 per cent nitric acid solution, followed by thorough washing in water. Grease or oil may be removed by the caustic

WELDED REPAIRS

soda solution or by using a mixture of $\frac{1}{2}$ carbon tetrachloride and $\frac{1}{2}$ gasoline. Leaded gasoline should not be used.

The flux should be removed from welds as soon as the weld is cold. Aluminum flux is corrosive to the metal and should be completely removed. The flux may be washed off with hot water and a scrub brush, or a jet of live steam may be directed along the welded area. Sometimes a light sandblast is used after treating with hot water or steam. Steam or hot water should be used after the sanding to remove any sand which might be stuck to the material. Welds in magnesium should never be sandblasted. Seams that are in corners and places hard to reach should be treated with a warm 10 per cent solution of dilute sulphuric acid. Treat the weld for 5 or 10 min. with the acid solution and thoroughly rinse with clean, cold water. All welds should be washed with clean, cold water to remove final traces of flux.

After welds are completely and thoroughly cleaned, they should be treated with a protective coating. The inside surface of hollow-tube members or parts which have open ends should be given a coat of oil-based primer. The coating is applied to the inside of the tubes by completely filling them with the oil-based primer and draining. In some cases the entire part may be dipped into the primer.

The inside surfaces of all closed or sealed steel parts or members, which are to be plated and which contain crevices or pockets where the protective solution might be held, should be protected by a coating of raw linseed oil or an approved rust-preventive compound. The protective material should be forced into the hollow members under pressure or by dipping the entire member in the liquid. The liquid should be held at a temperature of approximately 160° F., and remain in contact with the metal for not less than 2 or 3 min. After thoroughly draining, all excess oil should be wiped from the outside surfaces. The oil should be forced into the hollow members through a hole at the bottom in such amounts as to force the oil out of a hole at the top of the assembly. Holes drilled in members to inject the oil should be closed by self-tapping screws. These holes should never be drilled where they may weaken highly stressed parts. The outside of metal parts and surfaces should be treated with a coat of oil-based primer. The coat should be followed by two or more coats of oil enamel, which may be pigmented to match the original color. After treating with the oil-base primer, the outside surfaces of metal parts are usually coated with two or more coats of aluminized varnish. Aluminized varnish is prepared by stirring in 1 lb. of aluminum powder per gallon of the varnish.

After a joint has been silver-soldered or brazed, the scales formed by the flux should be removed. For removal of cold flux from copper and brass, a solution composed of 1 fl. oz. of sulphuric acid and $1\frac{1}{2}$ oz. of sodium dichromate to a gallon of water is used. The action of this solution is slow but will not injure the metal. After treatment with the solution, the welded joint should be cleaned by rinsing with water and dipped in a bright-dip solution which consists of the following: 20 fl. oz. of nitric acid, 68 fl. oz. of sulphuric acid, 0.12 fl. oz. of hydrochloric acid, and 40 fl. oz. of water. After the bright-dip treatment, which is used to restore the bright color to the parts, the parts should be thoroughly rinsed with clean running water. Flux and scale on steel parts after brazing, bronze welding, or soldering may be removed with a sand blast or wire brush. The flux may also be removed by dipping in boiling 10 per cent caustic soda solution for about 30 min. and rinsing thoroughly with clean water.

Some welds require finishing other than cleaning and the application of a protective covering. Such operations as grinding, polishing, and buffing may be required. It is necessary, at times, to remove the welding bead from both the top and bottom of the weld. This may be done by using the proper grinding wheel.

Where polishing and buffing of the seam is necessary, care should be taken that the weld is not ground down to such an extent that the final operation will leave a groove in the parent metal along the weld. The grinding operation should be stopped before the material to be removed is flush with the surrounding surface.

Polishing is similar to grinding, except that polishing is not designed to remove much material. Polishing removes small irregularities to obtain a smooth surface.

Buffing is necessary when a high polish is required. Soft wheels usually made from layers of cloth are used. Jewelers' rouge or similar material is applied to the cloth wheel to produce the desired finish.

Many of the aluminum alloys, after forming, are anodized before being used in aircraft structures. This is particularly true if the material is to be painted or have other finishes applied. Anodizing consists of forming an artificial oxidation on the surface of aluminum or magnesium alloys. The material to be anodized is submerged in a chemical solution and treated with an electric current. The surface thus formed resists corrosion and forms a good bond for paints or other

WELDED REPAIRS

finishes. Either a sulphuric acid or chromic acid electrolyte is used.

Testing Welded Joints. It is only by testing practice welds that the welding operator can discover his faults in welding. Each type of joint should be made and tested to discover various defects and the method of correcting them. The welding operator should not be satisfied until he can consistently weld each type of joint with a high degree of efficiency.

There are a number of methods of testing a weld to determine whether or not it has been properly made. There are six points which the welding operator can determine by visual inspection.

1. The weld should be as smooth and as nearly uniform in thickness as possible.

2. The weld should be examined for the necessary reinforcement.

3. The edge of the weld should taper off smoothly into the base metal, showing no distinct line.

4. Oxides formed on the base metal should not extend more than ½ in. from the weld.

5. The weld should show no indication of burns, distortion, cracking, or pitting.

6. There should be no signs of projecting or loose globules of metal, porous material, or blowholes.

After the welding operator has thoroughly examined the weld, it should be tested. Bend the piece of metal containing the weld into a U-shape. The top or bead of the weld should be on the inside of the U.

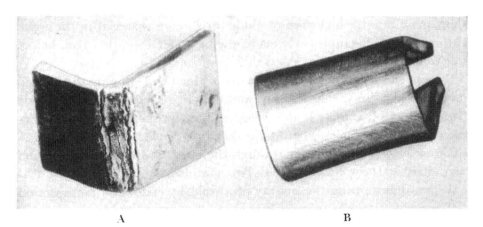

Fig. 263. Good and poor weld bent into U-shape. The poor weld "A" has failed. (Courtesy The Linde Air Products Company)

Place the U-shape in a vise and tighten it until the sides of the U are pressed closely together. The welded seam should project slightly above the jaws of the vise. If the break occurs in the weld itself, the weld has not been properly made. If breaking occurs along the edge of the weld, between the weld end and the base metal, proper fusion has not been

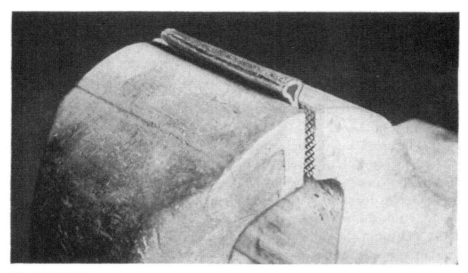

Fig. 264. Weld bent in vise. The welded seam has pulled away from the base metal. (Courtesy The Linde Air Products Company)

attained. The appearance of the broken part of a weld often indicates the fault in the welding. The break should be examined for crystalline structure. Burning due to an excess of oxygen is often indicated by coarse crystals which sometimes show peacock colors. Specks or small spots may indicate inclusion of oxides or foreign material in the weld. If the weld has been properly made, the base metal should break before the weld itself.

The strength of the weld may be determined by clamping the welded material in a vise at the weld line and then breaking the weld by blows from a heavy hammer. If the weld has been properly made and reinforced, the base metal should break along the edge of the weld. A hacksaw may be used to cut through the weld in order that the metal deposited in the weld itself may be examined.

When determining the quality of a weld by etching, a cross section of the weld at two or more points along the seam should be prepared. This may be done by cutting across the weld or by cutting lengthwise of the weld. Remove all saw marks from the surface to be etched with

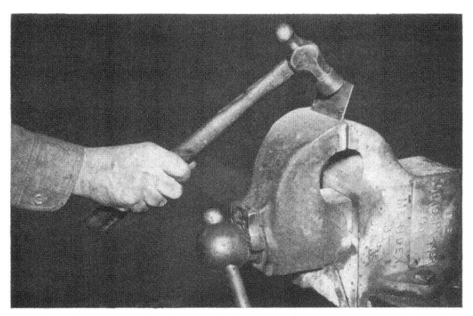

Fig. 265. Testing a weld by hammering in a vise. (Courtesy The Linde Air Products Company)

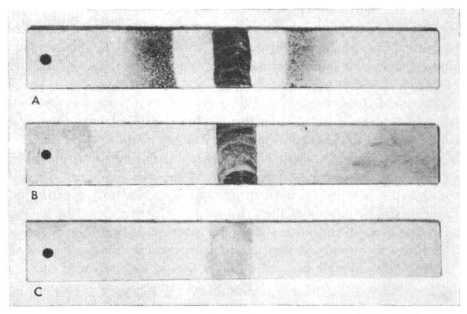

Fig. 266. Welds as they may appear after etching. (Courtesy The Linde Air Products Company)

a fine mill file. Polish out all file marks with emery cloth. Finish the surface to a mirror-like finish with fine emery cloth. Do not touch the polished surface with the hands or any substance which may leave a film of grease or oil. The exposed surface is then etched with a solution consisting of 1 part of nitric acid and 2 parts of water, or equal parts of hydrochloric acid and water. The nitric acid solution is the faster and attacks the metal immediately upon being applied. This fluid may be left on the metal for approximately 15 min., while the hydrochloric acid solution will require 1 or 2 hr. Apply the acid solution to the polished surface with a brush or swab and allow the piece to stand until all action has ceased. Wash the treated surface with water, followed by alcohol, and allow to dry thoroughly. Use a strong magnifying glass or microscope to examine the surface for defects. Uniformity of grain structure, depth of fusion, and any defects may be readily seen.

Rudder Construction and Repair. A typical rudder has a main vertical member of chrome-molybdenum steel airplane tubing, having an outside diameter of $7/8$ in. with a wall thickness of 0.031 in., and a trailing edge formed of light-wall tubing having an outside diameter of $3/8$ in. The main cross members are formed of $1/32$-in., 17S–T aluminum alloy. There are two steel piano-wire braces $3/64$ in. in diameter. The rudder horns to which the rudder cables are to be attached are formed from a single piece and are welded to the bottom of the vertical member. The hinges are made from chrome-molybdenum airplane tubing having an outside diameter of $1/2$ in. and an inside diameter of $3/8$ in., and they are $3/4$ in. long.

The main cross members and the steel-wire braces are fastened into place by brazing, using a Tobin Bronze brazing rod. The trailing-edge tube is welded to the vertical member. The top of the vertical member is closed by having a small plate of thin chrome-molybdenum sheet welded into the opening. The bottom of the main member is fitted through the hole in the horn fitting and is welded to the fitting on both sides. The bottom of the main member is then closed by having a circular piece of chrome-molybdenum sheet welded into the opening. The hinge fittings are welded to the main vertical member.

All welding operations on the rudder should be performed in a jig such as shown in Figure 235. To construct a jig, the rudder should be laid out full-sized on a sheet of plain manila paper. A $7/8$-in. strip of wood equal in length to the main vertical member should be fastened in place on the drawing which has been laid out on a large flat surface

WELDED REPAIRS

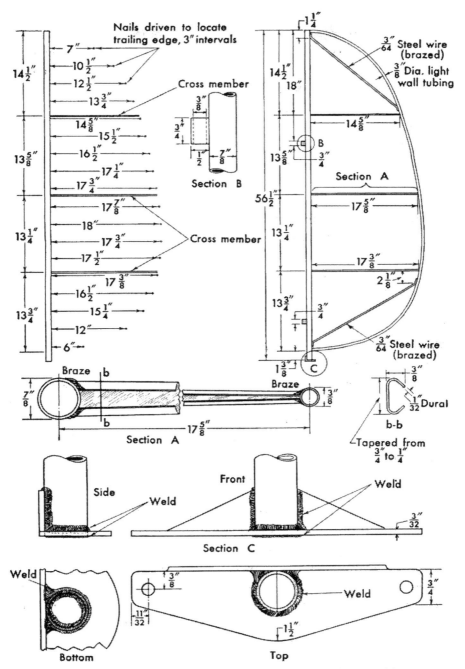

Fig. 267. Plan for layout and construction of welded frame for a rudder.

into which nails may be driven. Lay off on this drawing at 3-in. intervals along the strip representing the main member, the distances shown in Figure 266. At these points, drive nails just touching the inner edge of the trailing edge. A sheet of asbestos should be placed under each point where welding or brazing is to take place. Nails are driven at intervals along each side of the wood strip, which is then removed, and the main vertical member is fitted into place. The hinges should already have been welded to the vertical member and the member checked for straightness before being placed in the jig. Both ends of the trailing edge are then welded to the main member. The steel-wire braces are next brazed into place, and the main cross members fitted and brazed. The rudder can then be removed from the jig and the rudder horn, which has been formed from $3/32$ in. high-test nickel steel, is welded into place. Before welding the rudder horn into place it should be annealed, and after welding, it should be heated and quenched in oil to restore its hardness.

All of the welds should be carefully cleaned, and the inside of the tubes treated to prevent corrosion.

Vertical Fin Construction and Repair. Figure 268 shows the construction of a vertical fin. The main vertical member is of $3/4$-in. outside-diameter, chrome-molybdenum airplane tubing having a wall thickness of 0.063 in. The leading edge is of chrome-molybdenum airplane tubing having an outside diameter of $1/2$ in. and a wall thickness of 0.031 in. The cross braces, as shown in Figure 268, details A–A and B–B, are formed of 17S–T aluminum alloy. Detail C shows the construction of the upper rudder hinge and the insertion of two $1/4$-in. tubes through which wire braces pass.

A jig should be laid out as shown in Figure 268. The hinge fittings and the tubes through the main member should be welded into place, and the vertical member checked for straightness before placing in the jig. The top end of the leading edge should be welded to the vertical member. The opening in the top of the vertical member should be closed by welding in a small piece of thin chrome-molybdenum sheet. The bottom end of the vertical member is left open. The aluminum alloy cross braces are fastened into place by brazing, using a Tobin Bronze brazing rod.

After the fin is complete, all welds should be properly cleaned, and the inside of the tubes treated to prevent corrosion.

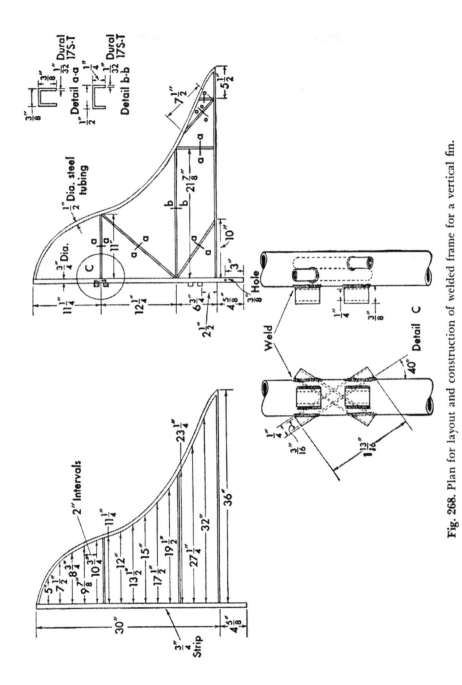

Fig. 268. Plan for layout and construction of welded frame for a vertical fin.

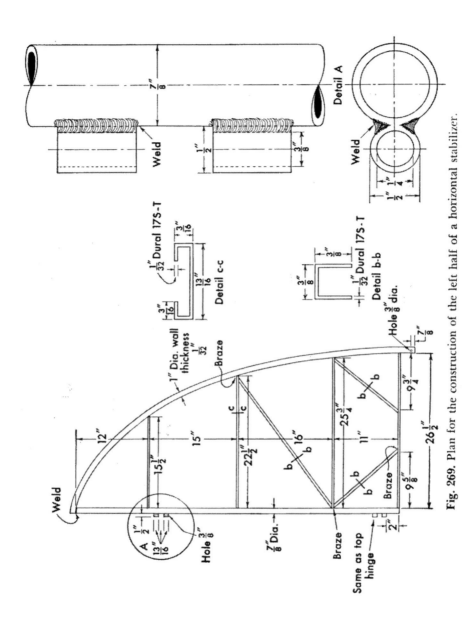

Fig. 269. Plan for the construction of the left half of a horizontal stabilizer.

WELDED REPAIRS

Stabilizer Construction and Repair. Figure 269 shows the construction of the left half of a horizontal stabilizer. The spar member to which the aileron hinges are fastened consists of $7/8$-in. outside-diameter, chrome-molybdenum airplane tubing, having a wall thickness of 0.063 in. The leading edge consists of chrome-molybdenum, thin-wall, airplane tubing having an outside diameter of 1 in. and a wall thickness of 0.031 in. The main cross members are formed of $1/32$ in., 17S–T aluminum alloy, as shown in Figure 269, detail C–C. The diagonal cross members are formed of the same material, as shown in detail B–B. Detail A shows the method of attaching the hinge fittings.

A suitable jig should be prepared. The hinge fittings should be welded to the vertical member, and the member checked for straightness before being placed in the jig. The top end of the leading edge should be welded to the spar member and the Dural braces brazed into place. The outside end of the main member should be closed by welding into the opening a piece of chrome-molybdenum sheet. The inner end is left open.

All welded and brazed joints should be properly cleaned, and the inside of the tubes treated to prevent corrosion.

Fuselage Construction and Repair. Figure 270 shows a perspective view of a section of a welded steel-tube fuselage. Figure 271 shows the bottom of this section and the front view of the front and rear truss members. Figure 272 shows the right side of the section which is identical to the left side. The top view of the section is also shown in Figure 272.

This section is constructed of chrome-molybdenum airplane tubing of the outside dimensions shown in the drawings, having a wall thickness of 0.063 in. The two sides of the fuselage should first be constructed in a suitable jig, after which the various cross members are welded into place, beginning with the members forming the front truss.

After welding is completed, all welded joints should be cleaned, and the inside of the tubes protected against corrosion.

Engine-Mount Construction and Repair. Figure 273 shows the side and rear view of a welded, steel-tube, engine mount. Figure 274 shows the top and bottom view of the same engine mount. Figure 275 shows a suggested jig to be used in the construction or welding in of parts of the engine mount shown in Figures 273 and 274.

The jig should first be constructed, as shown in Figure 275, using $1 3/4$-in. or 2-in. angle iron. It is important in the construction of the

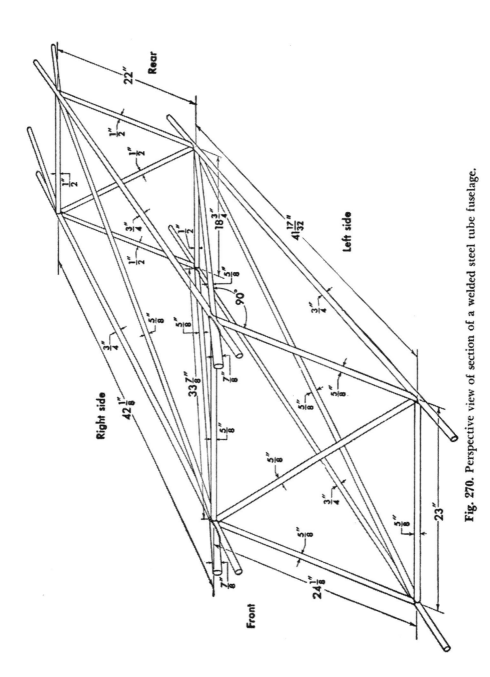

Fig. 270. Perspective view of section of a welded steel tube fuselage.

WELDED REPAIRS

jig that the holes for the engine and fuselage fittings be exactly located. Always check location before fastening.

The engine and fuselage fittings should first be constructed. Form the circular part from $3/32$-in. chrome-molybdenum sheet. These fittings are $1\frac{1}{4}$ in. in diameter and have fitted into them a chrome-molybdenum steel tube having an outside diameter of $\frac{1}{2}$ in. and an inside diameter of $3/8$ in. The tube is welded into the fitting, and the fittings bolted into place on the jig.

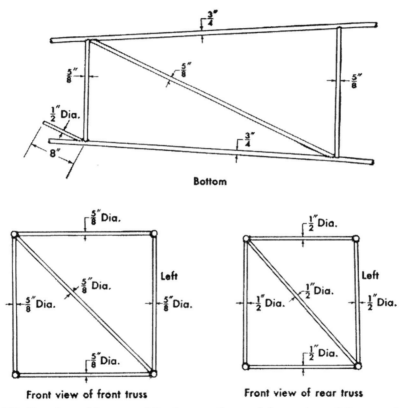

Fig. 271. Drawing of bottom of fuselage section and front view of front and rear trusses.

The horizontal members to which the engine is bolted are then constructed and welded into place on the engine fittings. It should be noted that the rear portion of these tubes and the main braces to the fuselage fittings are reinforced with an external sleeve. The two outside braces from the lower fuselage fittings are then fitted into place and welded, the rear members being welded into place first.

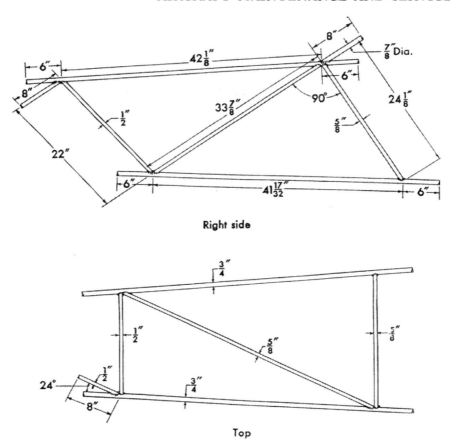

Fig. 272. Drawing of right side and top of fuselage section.

The upper braces connected with the upper fuselage fittings are fitted and welded into place. The rear diagonal brace is then welded into place, followed by the cross-front member. The last operation is the welding of the two gussets at the front of the engine mount.

All joints should be made in an approved manner. After thoroughly cleaning all welded joints, the inside of the tubes should be protected against corrosion.

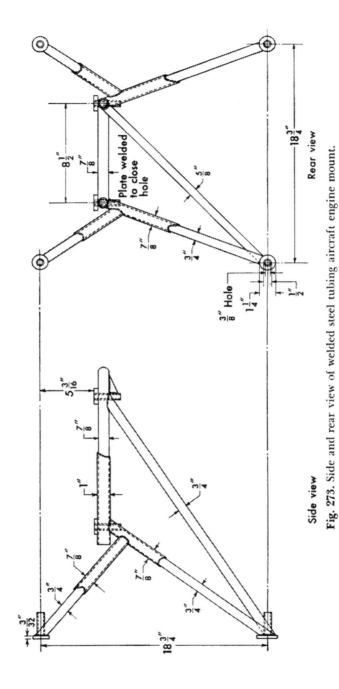

Fig. 273. Side and rear view of welded steel tubing aircraft engine mount.

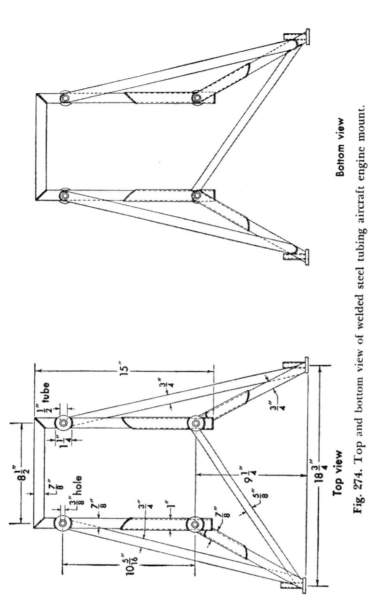

Fig. 274. Top and bottom view of welded steel tubing aircraft engine mount.

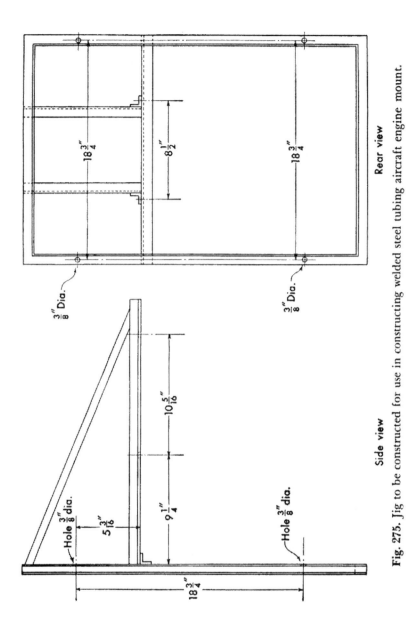

Fig. 275. Jig to be constructed for use in constructing welded steel tubing aircraft engine mount.

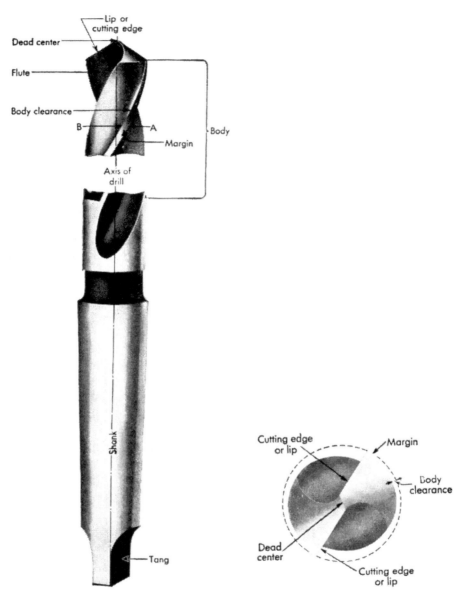

Fig. 276. Parts of a twist drill. (Courtesy Cleveland Twist Drill Company)

XIX DRILLING, BURRING, FILING, AND RIVETING IN SHEET METAL REPAIRS

Drilling. Most drilling in metal is done with a twist drill. As shown in Figure 276, the drill is made up of three principal parts: the point, the body, and the shank. Two spiral grooves run along opposite sides of the body. These grooves or flutes help form the cutting edges on the

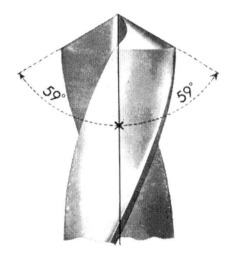

Fig. 277. Correctly ground lips. The two lips of this drill are of the same length and of the same angle to the axis of the drill. (Courtesy Cleveland Twist Drill Company)

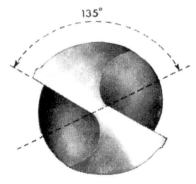

Fig. 278. How to gauge the correctness of lip clearance angle. (Courtesy Cleveland Twist Drill Company)

point. They also assist in removing the chips and allow lubricants to penetrate to the point of the drill. For general use, the point is usually ground to an angle of approximately 59°. This angle varies with the type of material being drilled.

The dead center is located at the center of the point and should be at the exact center of the drill, if the drill has been properly sharpened.

The point is the cone-shaped end of the drill. The edges of the flutes form the cutting part of the drill, or the lips. After sharpening, the lips should be equal in length, and the angle should increase somewhat to give lip clearance and allow the cutting lip to bite into the material. The angle of lip clearance should be gradually increased as the center of the drill is approached. The line across the dead center of the drill should form an angle of about 135°.

Fig. 279. Lip clearance. The drill on the left has no lip clearance. The drill on the right is properly cleared. (Courtesy Cleveland Twist Drill Company)

Improper grinding is a common cause for broken drills. On the edge of the flute is a small, narrow part which is equal to the diameter and allows proper clearance in the hole, preventing the drill from binding. This allowance is called body clearance. The metal column which separates the flutes is called the web and forms the main supporting column of the drill. The web thickens from the point to the shank.

The successful use of drills depends largely upon their proper grind-

Fig. 280. The web is the metal column which separates the flutes and is the supporting section of the drill. (Courtesy Cleveland Twist Drill Company)

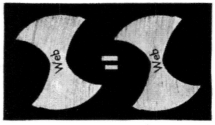

Fig. 281. Sectional view of the web. The section on the left was cut from a drill near the point, while the section on the right was cut near the shank. The difference in the thickness of the web at these two points is shown by the length of the white lines between the two sections in the illustration. (Courtesy Cleveland Twist Drill Company)

DRILLING, BURRING, FILING, AND RIVETING

ing. Two of the common faults in grinding drills are unequal length of the lips and an improper angle of lip clearance. The metal back of the lip itself should be ground away at an angle of about 12°. If this metal is not ground away, the lip has no clearance. Without clearance, the drill will not bite into the metal, but will slide over the surface

Fig. 282. A large drill press. Note shank of drill inserted in the chuck. (Courtesy Douglas Aircraft Company)

without cutting. Figure 283 shows proper lip clearance. Too much clearance will thin the lip to such an extent that the lip will chip. Chipping along the entire edge of the lip indicates too much lip clearance. Excessive speed may cause chipping of the lip toward the circumference of the drill, as shown in Figure 285. Figure 286 shows the effect of grinding one lip longer than the other or at a different angle. This will cause excessive strain on both the drill and the drill press. The hole will be uneven, scored, and larger than the diameter of the drill. The drill usually breaks at once.

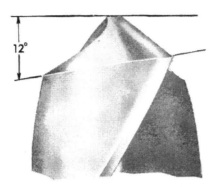

Fig. 283. The proper way to grind the surface back of the cutting lip. The angle indicated is at the circumference of the drill. This angle should be increased as the center of the drill is approached. (Courtesy Cleveland Twist Drill Company)

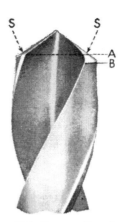

Fig. 284. Proper lip clearance. Note how much lower the heel line "B" is than the cutting lip line "A". This difference is the measure of clearance. (Courtesy Cleveland Twist Drill Company)

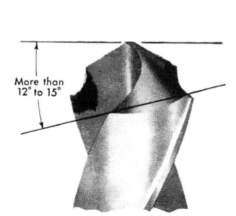

Fig. 285. Results of giving a drill too great lip clearance, or operating at an excessive speed. (Courtesy Cleveland Twist Drill Company)

Fig. 286. The angles of the lips are of different lengths. Note the effect on the hole. (Courtesy Cleveland Twist Drill Company)

DRILLING, BURRING, FILING, AND RIVETING

The feed of a drill is the distance that the drill moves down into the material with each revolution. If the feed is too rapid, the drill may bite in and break. In hard material, excessive feed may cause the drill to split up the center, as shown in Figure 287.

As the drill wears away, due to many sharpenings, it may be necessary to thin the web. This may be done by grinding it with a round-faced emery wheel or by grinding it with a flat, sharp-cornered emery wheel which produces a notched point, as shown in Figure 288.

Drill sizes are designated as follows: by numbers from 80 to 1, corresponding in drill sizes from 0.135 to 0.228; by the letters, A to Z, which correspond to the drill sizes, 0.234 to 0.413; and by fractions from $\frac{1}{64}$ in. to 4 in. and over, by sixty-fourths. Drill sizes are also given in the metric system in millimeters from 0.5 to 10 mm., by $\frac{1}{10}$ mm. Above 10 mm., the size increases by 0.5 mm. A millimeter is approximately $\frac{1}{25}$ in. and $\frac{1}{10}$ mm. is approximately $\frac{1}{250}$ in.

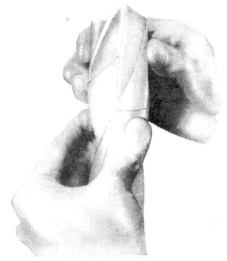

Fig. 287. This drill does not have enough lip clearance. As a result, cutting stopped and as the feed pressure increased, the drill "split up the center." (Courtesy Cleveland Twist Drill Company)

The speed of a drill is the speed of its circumference or the distance that the drill would roll over a flat surface. For example, if the drill

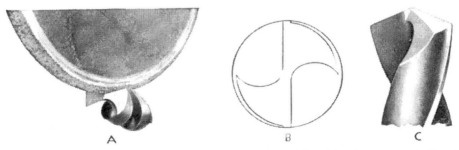

Fig. 288. Two methods of thinning the web of a drill. "A" using a round faced emery wheel; "B" the result of using a round-faced emery wheel; "C" a notched point formed by a sharp cornered emery wheel. (Courtesy Cleveland Twist Drill Company)

speed is 300 ft. per min., the drill would roll along a flat surface 300 ft. in one minute. The speed of a drill varies with the kind of material being drilled.

TABLE XI. SUGGESTED SPEEDS FOR HIGH-SPEED DRILLS

	Speed in F.P.M.
Aluminum and its alloys	200–300
Bakelite	100–150
Brass and bronze, ordinary	200–300
Bronze, high tensile	70–100
Cast iron, soft	100–150
Cast iron, hard	70–100
Cast iron, chilled	30–40
Malleable iron	80–90
Magnesium and its alloys	250–400
Monel metal	40–50
Steel machinery (0.2 to 0.3C)	80–110
Steel, annealed (0.4 to 0.5C)	60–70
Steel, tool (1.2C)	50–60
Steel forgings	50–60
Steel, alloy	50–70
Steel, stainless, free machining	60–70
Steel, stainless, hard	30–40
Slate, marble and stone	15–25
Wood	300–400

It is necessary at times to use a lubricant to assist in cutting. For brass and for aluminum and its alloys, gasoline and lead-oil compounds may be used. Cast iron is usually drilled dry; malleable iron, Monel metal, and steel may be drilled with a good nonviscous, neutral oil. When drilling hard material, turpentine is sometimes used to assist the lips in biting into the material.

The common sizes of rivets and the corresponding drill sizes are given in Table XII.

TABLE XII. COMMONLY USED RIVET AND DRILL SIZES

RIVET	DRILL
1/16	#51 (0.067)
3/32	#40 (0.098)
1/8	#30 (0.128)
5/32	#20 (0.161)
3/16	#10 (0.193)
1/4	F (0.257)
5/16	P (0.323)
3/8	W (0.386)

Burring. When sheet metal is sheared, punched, sawed, or drilled, the edge of the metal is usually left in a roughened condition. The sharp points left by the cutting tool are called "burrs." Tiny cracks may also appear on the cut edge. The removal of these burrs and cracks

DRILLING, BURRING, FILING, AND RIVETING

is called "burring." Proper burring leaves the edge of the metal smooth and even.

Burring is done for several reasons. Burrs may cause a material to crack while being formed. Cracks, too small to be seen by the naked eye, across the edge of a piece of sheet metal may cause minute cracks

Fig. 289. Burring stamped parts that are to be formed. (Courtesy Reynolds Metals Company)

to develop while the metal is being formed. These minute cracks will become enlarged and cause failure of the part in service. Burrs along the edge of parts may prevent a perfect fit. Burrs along the edge of pieces of sheet metal are almost sure to scratch and damage other sheets with which they come in contact. Serious cuts and scratches may be caused by burrs. Figure 290 shows the appearance of a magnified edge of a piece of sheet metal which has been cut. However, before burring a part, be sure that burring is necessary. Many production parts do not need to be burred.

In burring, not only the burrs on each edge of the sheet must be removed, but also enough of the edge of the sheet itself to remove any cracks which might have been formed. Any edge which is to be

AIRCRAFT MAINTENANCE AND SERVICE

bent or stretched must be burred. Edges that have been sawed are rough and ragged, but sawing does not usually cause cracks to appear in the edge of the sheet. Cracks may, however, start at the base of the

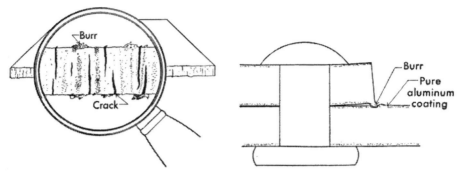

Fig. 290. Magnified edge of a piece of sheet metal that has been cut.

Fig. 291. Damage may be caused by burrs left on the edge of riveted sheets.

small grooves left by the saw teeth. Sawed material should have both the burrs removed and the edges smoothed. On drilled holes, it is usually only necessary to remove the rough burrs around the edge of the hole. Routed material should be burred and smoothed. Sheared material usually develops small cracks and should be burred and smoothed. Sawed, punched, or sheared edges are more likely to fail when forming than are rounded edges. Parts which have small toler-

Fig. 292. Hand burring a formed part. (Courtesy The Glenn L. Martin Company)

DRILLING, BURRING, FILING, AND RIVETING

ances must be carefully burred and the edges smoothed. Edges of sheet that are to be riveted must have all burrs removed.

Burring may be done by several different methods and by the use of various kinds of tools. Ordinary flat files may be used to remove burrs if carefully used. A fine, flat file is used to smooth the edge of the sheet if the edge is straight or nearly straight. A flat file may be used to smooth the edge of a sheet if the curve is convex. If the curve is concave, such as the inside of a hole, groove, or corner, files of various shapes may be used, such as half round, round, pippen, square, triangular, or knife. When burring a straight edge, the file should be pushed slantwise along the edge, and it should be pushed away from the workman.

When filing soft metals such as aluminum or Dural, allowing the file to drag lightly on the return stroke helps to keep it clean. In using

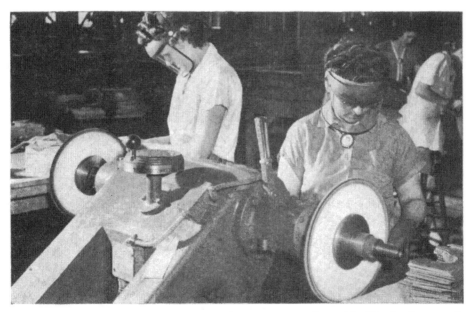

Fig. 293. Grinding wheels used for burring. (Courtesy Reynolds Metals Company)

a round or half-round file when smoothing a curve, the file should be rolled slightly in a clockwise direction as the stroke is made. This rolling motion allows the file to cut better and distributes the wear over all of the teeth. Sharp scrapers, such as hook scrapers or bearing scrapers, may be used for removing burrs. A V-edge scraper or a scraper containing small steel wheels similar to those in a kitchen knife sharpener may be used. Oversized drills, hand countersinks, or special burring tools are used to remove burrs from holes. In using rotary files or grinding wheels for burring, extreme care must be taken to remove only the desired amount of material, as power machinery cuts rapidly

Fig. 294. Removing burrs on a sander. (Courtesy The Glenn L. Martin Company)

and damage to the part may result. In production work, rotary sanders, steel-wire brushes, abrasive wheels, and steel rolls are used for burring and finishing the edges of cut sheet. Steel wool wrapped around a shaft attached to a motor may be used to burr and finish the edges of stamped or cut parts.

Filing. Filing is an operation which is used frequently in the working of metal. Files are of many types, and it is necessary to pick the right file for the job. All files should be equipped with a wooden handle while in use. The part of the file which fits into the handle is called the "tang." Figure 296 shows the various parts of a file. The file cuts in much the same way as a saw. On a single-cut file, the teeth extend across the face of the file in one direction. On a double-cut file, the teeth are cut across the face in two directions. A bastard file may be either single or double cut and has coarse teeth. A vixen file has curved teeth. A mill file has fine teeth and is a good all-purpose file for fine

DRILLING, BURRING, FILING, AND RIVETING

finishing. A Swiss file comes in various shapes and has fine teeth. Files also come in other shapes, such as flat, round, half-round, pippen (which is an oval shape), square, triangular, or knife edged. They should be kept clean with a file card which has a wire brush on one

Fig. 295. Removing burrs on a sander. (Courtesy The Glenn L. Martin Company)

side and a bristle brush on the other. The wire brush is used to loosen the particles in the file teeth, and the bristle to brush them away. Files should be wrapped in a cloth and separated from each other when put away. When using a file, usually the tip is held in the left hand and the handle of the file in the right hand. The strokes should be even,

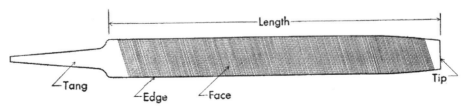

Fig. 296. The various parts of a file.

AIRCRAFT MAINTENANCE AND SERVICE

and the file should not be allowed to rock on the work if the surface is to be straight and level. Filing requires considerable practice in order to be done properly.

Riveting. Rivets are used to fasten the skin in place and to join many of the structural parts.

There are four main classifications of rivets, according to head shape: round head, brazier head, flat head, and countersunk or flush

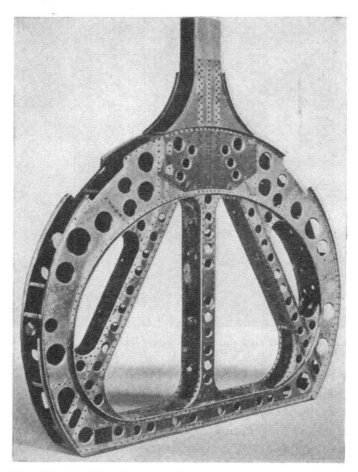

Fig. 297. A bulkhead showing large number of rivets. (Courtesy Aluminum Company of America)

head, as shown in Figure 298. The type of head used depends upon the part of the airplane that is being riveted. The countersunk head or flush type is used largely on exposed surfaces to reduce air resistance. Sometimes, on exposed surfaces, brazier-head rivets or shallow brazier-

DRILLING, BURRING, FILING, AND RIVETING

head rivets are used in place of the countersunk type. Rivets having round heads are used on structural parts. Flat-head rivets are used only on internal parts.

Rivets are also classified according to their alloy and whether or not they are driven "as received" or must be heat-treated. Rivets of 2S and 3S aluminum are quite soft and are driven cold, as received. Storing for long periods of time does not affect their driving characteristics or

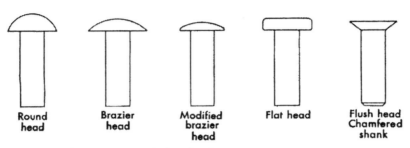

Fig. 298. Different types of rivet heads.

strength. Rivets of these alloys are used for nonstructural parts where great strength is not required.

Rivets of the alloy, 17S, are commonly used. These rivets are always driven in the fully heat-treated condition. If they are received in the 17S-O condition, they must be heat-treated and, if received in the 17S-T condition, they must be reheat-treated before driving. The rivets are heated to a temperature of approximately 940° F. and then quenched in cold water. The temperature must be accurately controlled, and these rivets should be used within 15 or 20 min. after heat treating. If not used immediately, they should be stored in a refrigerator to prevent hardening. This type of rivet is sometimes called an "icebox rivet." These rivets reach their full strength and hardness in about four days after driving.

Rivets of the alloy, 24S-T, have a higher strength than the 17S-T rivet and also require heat treating. They harden rapidly and should be driven within a few minutes after quenching. If not used immediately, they should be stored in dry ice which will keep them soft for a period of several weeks if held at a temperature of about — 50° F. This type of rivet is usually ordered in the fully heat-treated condition to guard against use in the annealed condition. After heat treating, if these rivets are allowed to age they are difficult to drive.

The alloy, 53S, in the W or T condition does not require heat treat-

AIRCRAFT MAINTENANCE AND SERVICE

ing in the aircraft factory. 53S–T is obtained by aging 53S–W for about six hours at 350° F. 53S–T may be stored indefinitely at room temperature before driving. 53S–W is easier to drive than the 53S–T, but does not reach its full strength and hardness until after several months.

The Army and Navy have set up specifications for rivets. The letters, AN, before a rivet number means that the rivet meets Army and Navy specifications. Rivets are designated by letters and figures, such as AN425–DD4–6. The AN signifies Army and Navy specification; 425 means that the rivet has a countersunk head and is of aluminum or aluminum alloy; DD indicates that the alloy is 24S–T; the 4 gives the diameter in thirty-seconds of an inch ($\frac{4}{32}$); and the 6 indicates the length in sixteenths of an inch ($\frac{6}{16}$). The numbers after the AN which indicate the head style are as follows:

420—An iron or copper rivet having a countersunk head.
425—An aluminum or aluminum alloy rivet having a countersunk head.
430—An aluminum or aluminum alloy rivet having a round head.
435—An iron or copper rivet having a round head.
442—An aluminum or aluminum alloy rivet having a flat head.
455—A brazier-head rivet.

If the letter A follows the number given above, it indicates 2S or 3S aluminum which should be driven as is without heat treating. The heads on these rivets are smooth. AD following the number indicates an A17S–T alloy to be driven as is without heat treating. This rivet has a dimple in the head. D indicates a 17S alloy, ice-box rivet

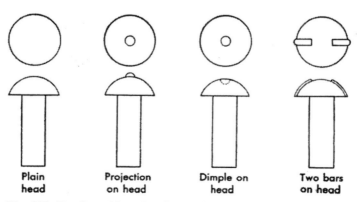

Fig. 299. Head markings for rivets of various alloys.

DRILLING, BURRING, FILING, AND RIVETING

which should be heat-treated before driving and has a small projection on the head. DD indicates a 24S alloy, ice-box rivet which must be heat-treated before driving. This rivet has two raised bars on the head. These markings are shown in Figure 299.

Rivet holes may be drilled or punched either before or after the

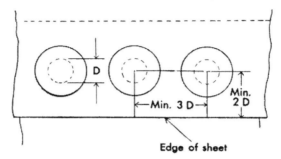

Fig. 300. Rivet spacing and edge distance.

part is in place. There should be just enough space between the rivet and the edge of the hole to allow the rivet to slip easily into place. The mechanic should be sure that he is using the proper-sized drill before drilling rivet holes. The most common sizes for aircraft rivets are $\frac{1}{16}$, $\frac{3}{32}$, $\frac{1}{8}$, $\frac{5}{32}$, and $\frac{3}{16}$. The corresponding drill sizes for these rivets are: 51, 40, 30, 20 and 10, respectively.

Rivets are usually removed by drilling through the head and driving out the rivet with a punch. A drill the exact size of the rivet should be used and it should be carefully centered on the rivet head. Do not drill past the surface of the sheet, as shown in Figure 301. The head is

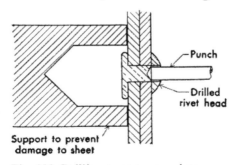

Fig. 301. Drilling to remove a rivet.

then either punched or sheared off with a cold chisel having rounded corners, and the rivet shank driven out of the hole with a suitable punch. Do not try to knock the head off a rivet with the rivet hammer. The driving of a rivet usually expands the rivet hole.

AIRCRAFT MAINTENANCE AND SERVICE

Sheet-Metal Repairs. Damage to sheet metal parts usually consists of cracks, dents, or small holes. If the damage is extensive, the entire sheet or sheets should be replaced. When replacing a sheet, the rivets should be of the same size and location as in the original structure. Cracks may be repaired by drilling a small hole at the end of the crack

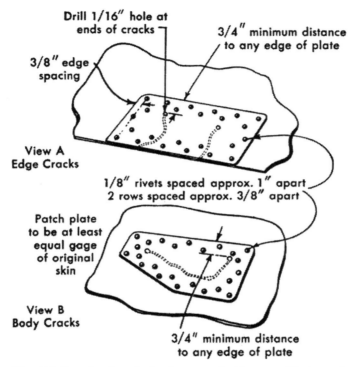

Fig. 302. Repair of cracks in sheet metal. (Courtesy Taylorcraft Aviation Corporation)

and then strengthening the part with a reinforcing patch of the same type of material. The patch should not be less than the thickness of the original material. The size of the patch depends on the size or length of the crack and, in no case, should the edge of the patch be closer than 3/4 in. to the crack. Several holes may be drilled through the damaged part and the patch, so that metal screws may be inserted to hold the patch in place while the rest of the rivet holes are drilled. These screws are replaced by rivets.

Dents in sheet-metal parts may be taken out by removing the part and placing the concave side of the dent on a smooth surface. The dent is then worked out by the use of a wooden or rawhide mallet. It is

DRILLING, BURRING, FILING, AND RIVETING

sometimes possible to remove a dent with the part in place by using a dolly or bucking bar on the concave side of the dent. Any damage to the finish should be repaired.

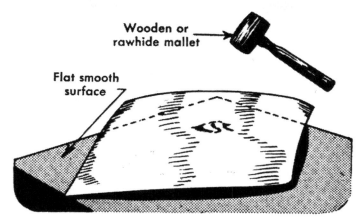

Fig. 303. Removing dents in sheet metal. (Courtesy Taylorcraft Aviation Corporation)

Small holes may be repaired by trimming out the ragged material and smoothing the edges of the holes with a pair of aviation snips or by the use of a file. The hole is then covered with a patch.

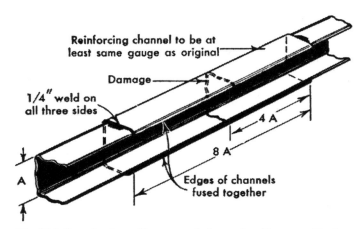

Fig. 304. Repair of small structural channels. (Courtesy Taylorcraft Aviation Corporation)

Structural parts, such as steel or aluminum alloy channels, may be repaired by proper reinforcement. The reinforcing material should be of the same type and thickness as the damaged member. If the reinforcement is to be used on the inside of the channel, it should be slightly thicker than the original material. Small structural channels that have been dented or bent should be carefully straightened. If cracks have occurred, a small hole should be drilled at each end of the

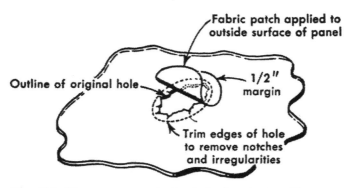

Fig. 305. Temporary repair for hole in plastic. (Courtesy Taylorcraft Aviation Corporation)

crack to prevent further extension. A reinforcing channel should then be cut and fitted snugly over the outside of the damaged section. The reinforcing channel should be the same thickness and the same material as the original.

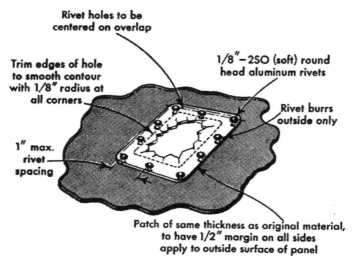

Fig. 306. Permanent repair for hole in plastic. (Courtesy Taylorcraft Aviation Corporation)

XX LAYOUT AND BEND ALLOWANCE

Layout. The technical problems involved in layout are those of the draftsman and engineer. However, it is necessary that the mechanic be able to use the fundamental layout tools. The workman should be able to square material to be cut, locate rivet holes, lay out material, and develop patterns, in order that he may cut the material to the required size. He should be able to use protractors, compasses, dividers, straightedges, trammel gauges, scribers, squares, and measuring instruments.

Layout, when it concerns the aircraft mechanic, consists largely of transferring information contained on a blue print to a paper pattern or to the material with which he is to work. All of the information necessary to complete a job is usually found on a blueprint, although the drawing on the blueprint is not usually full size. It is necessary that

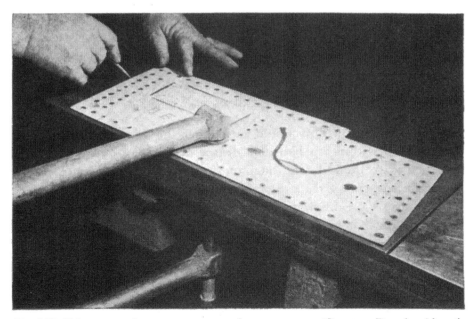

Fig. 307. Using a template as a pattern to lay out a part. (Courtesy Douglas Aircraft Company)

the workman be able to lay out a full-scale drawing. This layout may be made on the material itself, but if the job is in any way complicated, the layout should first be made on a sheet of heavy paper which may then be cut to the proper size and used as a pattern by which to cut the metal. When making a layout, the workman should use extreme care to check the pattern with the blueprint in every detail. The paper pattern should be fastened to the material with adhesive tape, and the outline drawn with a soft, sharp pencil. Care should be taken to hold the pencil in such a manner that the line closely follows the edge of the pattern. Most aircraft metals do not permit the use of a metal scriber. The scratches weaken the material and may lead to corrosion and the development of cracks. Holes to be drilled may be located by a center punch or prick punch, taking care not to damage the material. When the use of a punch is not allowed, the centers of the holes may be located by crossed pencil marks.

Tolerance and allowance are sometimes confused. Tolerance is the amount of variation permitted in the size of a part. Tolerance is usually indicated on the blueprint or on the template in the form of a plus or minus fraction. For example, the angle of a flange might be indi-

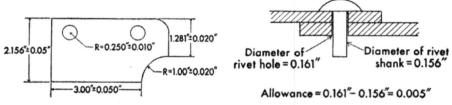

Fig. 308. Tolerance. **Fig. 309.** Allowance.

cated as $45° \pm \frac{1}{2}°$. This would mean that the workman must bend the angle to within one-half of a degree of the required angle. The length of a part might be indicated on the drawing as $25\frac{1}{2}$ in. $\pm \frac{1}{32}$ in. Most airplane parts have a tolerance of not more than $\frac{1}{32}$ in.

Allowance is the difference in dimensions between two parts to allow the proper fit. For example, the piston of an airplane engine does not have the same diameter as the cylinder, but there is an allowance for the expansion of the piston. A rivet hole is not exactly the same size as the rivet as an allowance is made so that the rivet may be put easily into place. Two sheets that join end to end do not fit exactly but an allowance is made for expansion and tolerance.

Bend Allowance. Most work done on sheet metal has to do with

LAYOUT AND BEND ALLOWANCE

changing its shape. This shaping is primarily bending. In order to avoid wasting material, it is important that the mechanic be able to determine exactly how much of the sheet will be required to complete the job.

Bend allowance, once thoroughly understood, becomes simple enough for the workman to use it easily and intelligently. A thorough understanding of the meaning of bend allowance should be had before attempting to work problems in bend allowance or in layout where bend allowance is involved. It is not enough to be able to use tables and formulas blindly.

Bend allowance is the amount of material which must be allowed to form the bend. This allowance is the amount of material actually needed to form the curved portion. When sheet metal is bent, the angle formed does not have a sharp edge. The vertex of the angle formed

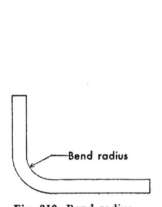

Fig. 310. Bend radius.

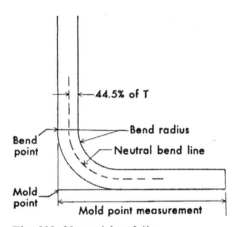

Fig. 311. Neutral bend line.

by the sides is not sharp, but follows the arc of a circle. The radius of this arc is called the "bend radius." The kind of metal and its thickness determine the bend radius. The bend radius is the radius of the arc forming the inner side of the bend, as shown in Figure 310. Bending material both stretches and shrinks the metal. The metal on the inside of the bend shrinks, or is squeezed together, while the metal on the outside of the bend is stretched. The line along which no stretching or shrinking takes place is called the "neutral line." This neutral line is not exactly halfway between the surfaces of the metal being bent. It is nearer the inside of the bend, being 44.5 per cent of the thickness of the metal from the inner surface, as shown in Figure 311. In laying out sheet metal to form an angle, the line forming the outside of the angle is extended until it meets the line forming the outside of the other side of the angle. These lines form the "flats" of the angle until they reach a point where the bend begins, as shown in Figure 312. The line from this point to the vertex of the angle thus formed is called the "mold line." The distance from the point of bend to the vertex of this angle is called "setback." In an angle of 90°, this setback is equal to the bend radius plus the thickness of the metal. This statement is written as $R + T$.

From Figure 312 it is readily seen that the amount of material required to form the angle is equal to the distance from A to B, this

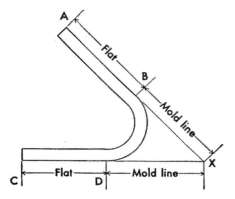

Fig. 312. Flats and mold line.

being the flat of one side, plus the distance from C to D, which is the flat of the other side, plus the material forming the bend itself. The length of AX is the flat of one side plus the mold line, the mold line being the distance from B to X. The distance from B to X is equal to the length of the flat on the other side plus the mold line from C to X. These are the distances which are usually given in laying out the angle. If the distance, AX, is equal to $1\frac{1}{2}$ in., and the distance, BX, is equal to 2 in., it might be easy to assume that the overall length, or developed flat pattern for the material needed, would be the sum of these two measurements, or $3\frac{1}{2}$ in. If a strip of metal $3\frac{1}{2}$ in. long were bent to form a 90° angle, it would be found that the sides of the angle exceeded the desired measurements given on the drawing. This is because the metal does not extend to the vertex of the angle but cuts across the corner, requiring a smaller amount of metal than that required to form a 90° angle.

LAYOUT AND BEND ALLOWANCE

The circumference of a circle is less than the length of the sides of a square having the same radius. The amount of metal required to form the bend itself is always less than the sum of the two mold lines forming the angle, as shown in Figure 312. The amount of material necessary to form the bend may be figured by finding the portion of the circumference of a circle equal in degrees to the angle formed by the sheet metal. The radius of this circle would be equal to the bend radius plus 44.5 per cent of the thickness of the metal. For example, if a 90° bend were to be formed in material 0.032 in. thick having a bend radius of ⅜ in., the amount of material necessary to form the bend could be found as follows: The circumference of the neutral circle may be found by using the formula $2\pi R$; π is equal to 3.1416. To find the radius of the neutral circle, multiply the thickness of the metal (0.032 in.) by 0.445 (the decimal equivalent of 44.5 per cent) and add this answer to the bend radius, which is 0.375 (decimal equivalent of ⅜); 44.5 per cent of the thickness of the metal is 0.014; 0.375 plus 0.014 gives 0.389, which is the radius of the neutral circle; 0.389 times 2 times π is 2.444, which is the circumference of the neutral circle to the nearest one-thousandth of an inch. Since only ¼ of the entire circumference is needed to form the bend, this number is divided by 4, giving 0.611 in. which is the amount of metal needed to form the bend itself.

If the angle had been formed through 45°, the amount of material needed to form the bend would be ½ that above. To form a complete circle, the amount of material needed would have been equal to the circumference of the neutral circle, or 2.444. To form an angle of 1°, the circumference of the neutral circle, which is 2.44416, is divided by 360 (the number of degrees in a circumference), giving 0.00679. This is the amount of material needed for each degree of bend when the radius is ⅜ in. and the thickness of the material is 0.032.

To save the above figuring each time an angle is laid out, tables have been prepared showing the amount of material or bend allowance for each degree of bend, or the amount of material needed for a 90° bend in materials of various thicknesses and different bend radii.

In the lefthand column of Table XIII look up ⅜ in., which is the bend radius; then find, in the top row of figures, the thickness of the material, 0.032. Where these two columns cross each other, we come to the figure 0.00679, which is the amount of material found necessary in the preceding problem to form a 1° bend.

TABLE XIII. TABLE FOR FINDING AMOUNT OF MATERIAL FOR EACH DEGREE OF BEND

R \ T	.010	.012	.015	.018	.020	.022	.025	.028	(1/32) .031	.032	.035	.040	.045	.049	.050	.051	.057	.062	.064	.065
1/64	.00035	.00037	.00039	.00041	.00043	.00044	.00047	.00049	.00051	.00052	.00055	.00058	.00062	.00065	.00066	.00067	.00072	.00076	.00077	.00078
1/32	.00062	.00064	.00066	.00068	.00070	.00072	.00074	.00076	.00079	.00079	.00082	.00086	.00090	.00093	.00093	.00094	.00099	.00103	.00104	.00105
3/64	.00089	.00091	.00093	.00096	.00097	.00099	.00101	.00103	.00106	.00107	.00109	.00113	.00117	.00120	.00121	.00121	.00126	.00130	.00132	.00132
1/16	.00117	.00118	.00121	.00123	.00125	.00126	.00128	.00131	.00133	.00134	.00136	.00140	.00144	.00147	.00148	.00149	.00153	.00157	.00159	.00160
5/64	.00144	.00145	.00148	.00150	.00152	.00153	.00156	.00158	.00160	.00161	.00163	.00167	.00171	.00174	.00175	.00176	.00181	.00184	.00186	.00187
3/32	.00171	.00173	.00175	.00177	.00179	.00181	.00183	.00185	.00188	.00188	.00191	.00195	.00199	.00202	.00202	.00203	.00208	.00212	.00213	.00214
7/64	.00198	.00200	.00202	.00204	.00206	.00208	.00210	.00212	.00215	.00216	.00218	.00222	.00226	.00229	.00230	.00230	.00235	.00239	.00241	.00241
1/8	.00226	.00227	.00230	.00232	.00233	.00235	.00237	.00239	.00242	.00243	.00245	.00249	.00253	.00256	.00257	.00258	.00262	.00266	.00268	.00269
9/64	.00253	.00254	.00257	.00259	.00260	.00262	.00265	.00267	.00269	.00270	.00272	.00276	.00280	.00283	.00284	.00285	.00290	.00293	.00295	.00296
5/32	.00280	.00282	.00284	.00286	.00287	.00290	.00292	.00294	.00296	.00297	.00300	.00304	.00307	.00311	.00311	.00312	.00317	.00321	.00322	.00323
11/64	.00307	.00309	.00311	.00313	.00314	.00317	.00319	.00321	.00324	.00324	.00327	.00331	.00335	.00338	.00339	.00339	.00344	.00348	.00349	.00350
3/16	.00335	.00336	.00339	.00341	.00342	.00344	.00346	.00348	.00351	.00352	.00354	.00358	.00362	.00365	.00366	.00367	.00371	.00375	.00377	.00378
13/64	.00362	.00363	.00366	.00368	.00370	.00371	.00374	.00376	.00378	.00379	.00381	.00385	.00389	.00392	.00393	.00394	.00398	.00402	.00404	.00405
7/32	.00389	.00391	.00393	.00395	.00396	.00398	.00401	.00403	.00405	.00406	.00409	.00412	.00416	.00419	.00420	.00421	.00426	.00430	.00431	.00432
15/64	.00416	.00418	.00420	.00422	.00423	.00426	.00428	.00430	.00433	.00433	.00436	.00440	.00444	.00447	.00448	.00448	.00453	.00457	.00458	.00459
1/4	.00443	.00445	.00447	.00450	.00451	.00453	.00455	.00457	.00460	.00461	.00463	.00467	.00471	.00474	.00475	.00475	.00480	.00484	.00486	.00486
9/32	.00498	.00500	.00502	.00504	.00505	.00507	.00510	.00512	.00514	.00515	.00517	.00521	.00525	.00528	.00529	.00530	.00535	.00539	.00540	.00541
5/16	.00552	.00554	.00556	.00558	.00559	.00562	.00564	.00566	.00569	.00570	.00572	.00576	.00580	.00583	.00584	.00584	.00589	.00593	.00595	.00595
11/32	.00607	.00608	.00611	.00613	.00614	.00616	.00619	.00621	.00623	.00624	.00626	.00630	.00634	.00637	.00638	.00639	.00644	.00647	.00649	.00650
3/8	.00661	.00663	.00665	.00667	.00668	.00671	.00673	.00675	.00678	.00679	.00681	.00685	.00689	.00692	.00693	.00693	.00698	.00702	.00704	.00704
13/32	.00716	.00717	.00720	.00722	.00723	.00725	.00728	.00730	.00732	.00733	.00735	.00739	.00743	.00746	.00747	.00748	.00752	.00756	.00758	.00759
7/16	.00770	.00772	.00774	.00776	.00777	.00780	.00782	.00784	.00787	.00788	.00790	.00794	.00798	.00801	.00802	.00802	.00807	.00811	.00812	.00813
15/32	.00825	.00826	.00829	.00831	.00832	.00834	.00837	.00839	.00841	.00842	.00844	.00848	.00852	.00855	.00856	.00857	.00861	.00865	.00867	.00868
1/2	.00879	.00881	.00883	.00885	.00886	.00889	.00891	.00893	.00896	.00896	.00899	.00903	.00907	.00910	.00911	.00911	.00916	.00920	.00921	.00922
17/32	.00934	.00935	.00938	.00940	.00941	.00943	.00945	.00947	.00950	.00951	.00953	.00957	.00961	.00964	.00965	.00966	.00970	.00974	.00976	.00977
9/16	.00988	.00990	.00992	.00994	.00995	.00998	.01000	.01002	.01005	.01005	.01008	.01012	.01016	.01019	.01019	.01020	.01025	.01029	.01030	.01031
19/32	.01043	.01044	.01047	.01049	.01050	.01052	.01054	.01057	.01060	.01060	.01062	.01066	.01070	.01073	.01074	.01075	.01079	.01083	.01085	.01086

Bend allowance chart, for any degree of angle of bend. (Courtesy of The Glenn L. Martin Company)

LAYOUT AND BEND ALLOWANCE

When the mechanic is thoroughly familiar with the above, he may approach the next step in figuring bend allowance which is finding setback, so that he may determine accurately the length of the flats forming each side of the angle. The setback is the distance from the vertex of the angle, formed by the mold lines, to the point where the bend begins on each side of the angle.

To find the setback an empirical formula is used. An empirical formula is one which has been worked out by experiment. The setback is found by adding the thickness to the bend radius $(T + R)$ and multiplying this number by the cotangent of one half the angle formed by the sides, as shown in Figure 313. For example, the angle formed by the sides in Figure 313 is 60°. From a table of natural

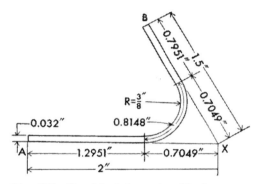

Fig. 313. Empirical formula—60 degrees.

cotangents, it is found that the cotangent of 30°, which is one half the angle formed by the sides, is 1.7320. If the bend radius is ⅜ in., and the thickness of the material is 0.032 in., the formula becomes:

$$(T + R) \cot \frac{\text{angle}}{2} = (0.032 + 0.375)\, 1.7320 = 0.7049$$

This number represents the setback which must be subtracted from the distance from A to X and from B to X to determine the length of the flat on each side of the angle. Since the distance from A to X is 2 in., and from B to X is 1½ in., the flat, A to C, is equal to $2 - 0.7049 = 1.2951$. In a similar manner, the flat forming B to D is equal to $1.5 - 0.7049 = 0.7951$. It can now be seen that the amount of material needed to form the given angle is 1.2951, which is the distance from A to C plus 0.7951, which is the distance from B to D plus the bend allowance. Since the angle is equal to 60°, we find from the table for 1° bends the amount of material needed when the thickness is 0.032

and the bend radius is 3/8 in. to be 0.00679. This number multiplied by 120, since the metal is bent through 120° to form a 60° angle, is 0.8148, which is the amount of material needed to form the bend. Adding 1.2951, 0.7951, and 0.8148, it is found that the required amount of material to form the given angle is 2.9050. This is called the developed flat pattern, as shown in Figure 314.

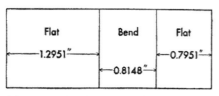

Fig. 314. The developed flat pattern.

Another empirical formula, instead of using the cotangent of one half the angle formed between the sides, uses the tangent (tan) of one half the angle through which the metal has been bent. The values obtained would be the same because each angle is the supplement of the other. The cotangent of an angle is equal to the tangent of its supplementary angle. Supplementary angles are two angles whose sum is equal to 180°.

A practical shop bend allowance formula, as given below, is often used.

Bend allowance formula:

$$[(0.01743 \times R) + (0.0078 \times T)] \times \text{no. degrees of bend.}$$

The procedure for using the above formula is as follows.

1. When the inside dimensions of a bend are given, subtract one radius for each dimension.

2. When the outside dimensions of a bend are given, subtract one radius and one thickness for each dimension.

3. When there is a bend on each end of a dimension, the following procedure is followed. When the dimension is given outside, subtract one radius and one thickness for each bend, or a total of two radii and two thicknesses. When the inside dimensions are given, subtract only the two radii.

Setback charts have been designed, such as the one shown in Figure 315, to determine the setback without the use of the empirical formula illustrated in the preceding problem. In using the setback chart, a straightedge is laid across the chart from top to bottom connecting the figures at the top, representing the bend radius, with the proper figure at the bottom which shows the thickness of the material. The words "open" and "closed" will be noted along the right side of

LAYOUT AND BEND ALLOWANCE

the chart. These words tell whether the angle is less than 90° or greater than 90°. If an angle is greater than 90°, it is called an open angle, while one less than 90° is called a closed angle. The rows of figures above and below 90° represent the number of degrees an angle is

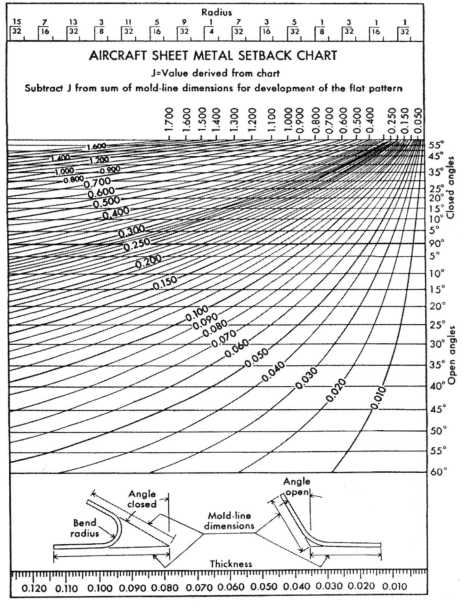

Fig. 315. Set back chart.

AIRCRAFT MAINTENANCE AND SERVICE

greater than or less than 90°. Ninety degrees is represented by the heavy line near the center of the chart and parallel to the bottom of the chart. An angle of 80°, which is 10° less than 90° and therefore a closed angle, would be represented by the number, 10, above the 90° point. While an angle of 120°, which is 30° greater than 90° and therefore an open angle, would be represented by the figure, 30, below the 90° point.

If the bend radius is ⅜ in. and the thickness is 0.032, the straightedge should connect these points on the chart. For an angle of 60°, the line opposite 30, which is above the 90° line since it is a closed angle, is located and followed across the chart to the point where it intersects the straightedge. This point falls between the line marked 0.600 and that marked 0.700, and is the required setback, or J value. Estimating this value since the point does not fall on a given line, the setback is found to be approximately 0.631. This amount is to be subtracted from the sum of the mold lines of the angle. The sum of the mold lines of the angle, as shown in Figure 316, is 3½, from which is subtracted 0.631, leaving 2.869, which is the developed length of the material necessary to form the flange. The setback chart is not as accurate as the empirical formula, but it is sufficiently accurate for most aircraft, sheet-metal work. The developed length will be within a few one-hundredths of an inch and usually falls within the tolerance limits.

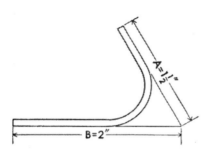

Fig. 316. An angle to show the use of set back.

Another type of chart, as shown in Figure 317, is also used to figure bend allowance. The first column, T, shows the thickness of the material. In the second column, J, is found the setback or J value for 90° bends. The values found on the chart are equal to twice the values to be used for a 90° bend. The third column, 1°, gives the value for the bend allowance for 1° for different combinations of bend radii and thickness. The fourth column, R, gives the bend radii.

To use this chart, a straightedge is placed across the chart from column T to column R, connecting the thickness of the material with the desired bend radius. Read the value in column J where the straightedge crosses the vertical line in the column to get a value which is double that for a 90° bend. For a single 90° bend, this value divided

LAYOUT AND BEND ALLOWANCE

by 2 and subtracted from the sum of the sides of the angle, including the mold line, gives the approximate developed length of the flat pattern. Figure 317A, B, C, D, and E illustrates the use of this part of the table. To find the bend allowance for each degree of bend, the

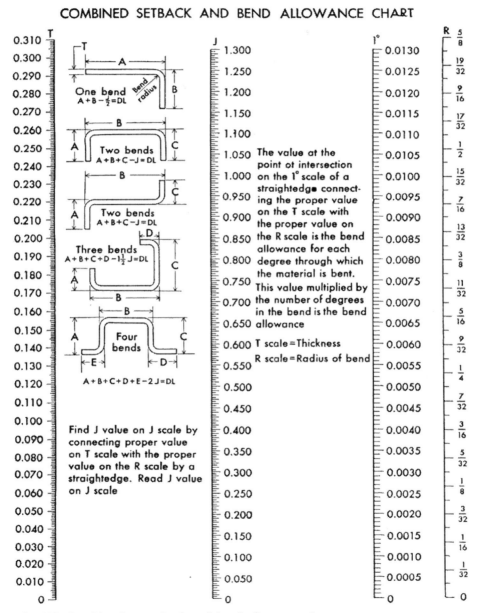

Fig. 317. Combination set back and bend allowance chart.

AIRCRAFT MAINTENANCE AND SERVICE

straightedge is placed across the chart connecting the thickness with the desired bend radius, and the number is read from the vertical line in the column, 1°. This number is then multiplied by the number of degrees through which the metal is bent to find the amount of material necessary to make the bend. This number is added to the sums of the flats forming the sides of the angle to develop the flat pattern.

Figure 318 shows how to use Table XIV in figuring the bend allowance from the empirical formula.

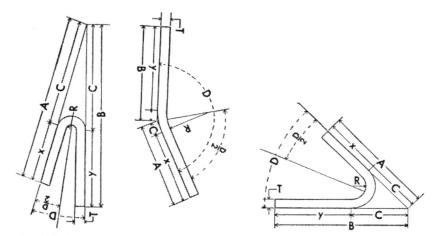

Fig. 318. Examples for use of Table XIV.

TABLE XIV. TABLE TO FIND DEVELOPED LENGTH OF 90° BENDS

R' / T	1/32	3/64	1/16	5/64	3/32	7/64	1/8	9/64	5/32	11/64	3/16	13/64	7/32	15/64	1/4	9/32	5/16	11/32	3/8	7/16	1/2
.330															.422	.466	.499	.512	.525	.549	.579
.284															.396	.435	.479	.492	.506	.529	.557
.259														.390	.374	.410	.449	.462	.476	.499	.529
.238													.361	.368	.354	.388	.423	.437	.450	.473	.504
.220												.333	.340	.347	.326	.367	.401	.415	.428	.451	.482
.203											.299	.306	.313	.319	.308	.339	.381	.394	.408	.431	.461
.180										.274	.281	.288	.294	.301	.287	.312	.353	.366	.379	.403	.433
.165									.247	.254	.260	.267	.273	.280	.270	.300	.335	.348	.361	.385	.415
.148								.223	.230	.237	.243	.250	.257	.263	.253	.283	.314	.327	.340	.364	.394
.134							.199	.206	.213	.220	.226	.233	.240	.246	.240	.266	.297	.310	.324	.347	.377
.120						.179	.186	.193	.199	.206	.213	.220	.226	.233	.223	.253	.280	.293	.307	.330	.360
.109					.156	.162	.169	.176	.182	.189	.200	.203	.204	.216	.208	.236	.263	.276	.243	.317	.347
.095				.135	.141	.148	.154	.161	.168	.175	.181	.188	.195	.201	.195	.222	.250	.263	.276	.300	.330
.083			.144	.121	.128	.134	.141	.148	.155	.161	.172	.175	.181	.188	.186	.208	.235	.248	.262	.285	.315
.072		.100	.106	.113	.119	.126	.133	.139	.146	.153	.164	.166	.173	.180	.178	.200	.222	.235	.248	.272	.302
.065		.092	.098	.104	.111	.117	.124	.131	.138	.144	.155	.158	.164	.171	.167	.191	.213	.226	.240	.263	.293
.058	.084	.081	.087	.093	.100	.107	.113	.120	.127	.133	.144	.147	.153	.160	.158	.180	.205	.218	.231	.255	.285
.049	.073	.072	.078	.084	.091	.098	.105	.111	.118	.125	.136	.138	.145	.152	.150	.172	.194	.207	.220	.244	.274
.042	.065	.064	.070	.076	.083	.090	.096	.103	.110	.116	.127	.130	.136	.143	.146	.163	.185	.198	.212	.235	.266
.035	.056	.060	.066	.072	.079	.086	.093	.099	.106	.113	.119	.126	.133	.139	.141	.160	.177	.190	.203	.227	.257
.032	.053	.055	.061	.068	.074	.081	.085	.094	.101	.108	.114	.121	.128	.135	.138	.155	.173	.186	.200	.223	.253
.028	.048	.052	.057	.064	.071	.077	.085	.091	.097	.104	.111	.118	.124	.131	.134	.151	.168	.182	.195	.218	.249
.025	.044	.048	.054	.060	.067	.074	.080	.087	.094	.100	.107	.114	.121	.127	.134	.147	.164	.178	.191	.214	.245
.022	.040	.045	.051	.058	.065	.071	.078	.085	.091	.098	.104	.111	.118	.125	.132	.145	.161	.174	.188	.211	.241
.020	.038	.045	.051	.058	.065	.071	.078	.085	.091	.098	.104	.111	.118	.125	.132	.145	.158	.172	.185	.208	.239
.018	.035	.043	.049	.055	.062	.069	.076	.082	.089	.096	.102	.109	.116	.122	.129	.143	.156	.169	.183	.206	.236

Table for finding the developed length of 90° bends. Subtract the correct figure in the table from the sum of the length of the legs. (Courtesy of The Glenn L. Martin Company)

XXI FORMING SHEET METAL

Shrinking and Stretching. Cold-working on aluminum alloys brings about strain hardening. A part may become brittle due to extensive cold-working. Metals and alloys in the annealed condition will stand more working than when in the T condition. Most parts should be worked, if possible, in the annealed condition. If the handworking operation is to remove distortion caused by heat treating, the work should take place immediately after heat treating before age-hardening conditions have developed.

Fig. 319. Forming aircraft skin on a stretching machine. (Courtesy The Glenn L. Martin Company)

FORMING SHEET METAL

The term, stretching, when used in connection with hand forming means that the material lengthens due to the operations performed upon it. Pure aluminum and alloys in the annealed condition are soft and malleable. The stretching action does not seem to weaken the material if it is not carried to excess. In stretching, the metal becomes thinner, just as a rubber band does when it is stretched.

Shrinking metal is just the opposite of stretching. The shrinking also causes hardening and sets up strains in the metal. Instead of the metal being thinned and lengthened as in stretching, shrinking causes the metal to become thicker and to decrease in area. The shrinking process is similar to the upsetting of a rivet. Figure 320 shows the rivet

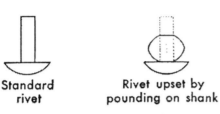

Fig. 320. A rivet before and after upsetting.

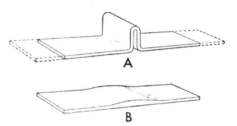

Fig. 321. The principle of shrinking: (A) A strip of metal is bent into a fold. (B) The fold has been completely hammered out, leaving the sheet shorter and thicker.

before and after upsetting. The rivet contains the same amount of material but is shorter and thicker than it was before the upsetting took place. If a strip of metal is bent as shown in Figure 321A and the bend hammered out by striking at the top of the fold, the metal in the sides of the fold is upset in a way similar to the upsetting of the metal in the rivet. As the fold is hammered completely out, as shown in Figure 321B, the strip becomes shortened, but is considerably thicker at the point of hammering. Severe bends of this type are not practical, but the illustration is used to show the principle of shrinking.

The V-block may also be used to shrink one side of a flange, as shown in Figure 322. This process is the opposite of that used to stretch one side of the flange. When shrinking one side of a flange in this manner, the flange will have a tendency to close. When the flange is opened, the tendency is to straighten the curve which has been formed in the side of the flange. The curved part should be clamped between two blocks while the flange is opened.

AIRCRAFT MAINTENANCE AND SERVICE

A shrinking block made of steel or other suitable material, as shown in Figure 323, is useful in shrinking flanged parts. The crimps placed in the edge of the part to be shrunk are clamped in the shrinking block

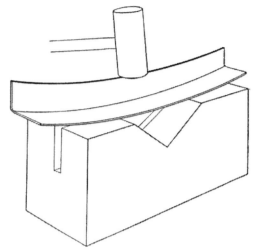

Fig. 322. A "V" block used to shrink one side of a flange.

and hammered out, as shown in Figure 324. Flanges may be made by the use of a proper form block and backing block, as shown in Figure 325. The blocks with the material to be formed between them are clamped in a vise. Using the proper mallet, the material is bent into

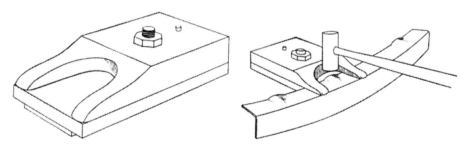

Fig. 323. A shrinking block.

Fig. 324. Using a shrinking block to shrink one side of a flange.

place over the form block. The form block may be made from hardwood, Masonite, Micarta, plastic, or suitable metal. The form block should have the proper radius on the edge over which the metal is to be formed, and a sufficient allowance should be made for spring back. The metal should be formed over the block by use of a cross-peen

FORMING SHEET METAL

wooden mallet, and the blows struck close to the form block, as shown in Figure 326. Curved flanges which are difficult to form may be made in this manner. In forming a curved flange the metal must be shrunk.

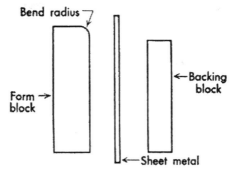

Fig. 325. A form block, the material to be formed and a backing block.

The excess metal should be supported by a backing-up or shrinking tool to prevent sharp wrinkles from forming, as shown in Figure 327. This tool also assists in holding the metal tightly to the form block.

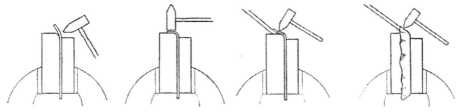

Fig. 326. Forming sheet metal by use of a form block, backing block, and wooden mallet.

Fig. 327. Using a backing-up tool to prevent wrinkles when forming a curved flange.

Forming Aluminum and Aluminum Alloys. Aluminum is one of the most workable of the metals used in aircraft fabrication. The workability of a metal is that property which allows it to be bent, formed, drawn, or spun without developing cracks or becoming weakened. Unlike the ferrous metals, aluminum and its alloys do not appear to have a definite yield point. However, there is a point beyond which the force necessary to deform the aluminum is not in direct proportion to the amount of deformation. The aircraft mechanic should not think of aluminum as being a simple material, but should realize that there are a large number of aluminum alloys having different forming characteristics due to their difference in hardness, springiness, and

Fig. 328. Hand finishing machine-formed parts distorted by heat treatment. Note distorted parts in background. (Courtesy The Glenn L. Martin Company)

Fig. 329. Hand finish forming a nose piece using a wood tool and mallet. (Courtesy The Glenn L. Martin Company)

FORMING SHEET METAL

ductility. The unit force used in forming aluminum has been defined as that force which produces a permanent set of 2 per cent of the original dimensions. This force is defined as the yield force or yield strength of the material.

When aluminum is rolled, a grain structure is set up similar to the grain in a piece of wood. Like a piece of wood, the aluminum sheet is

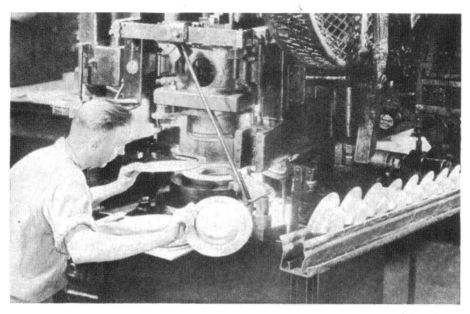

Fig. 330. The first drawing operation on a deep drawn part. (Courtesy Aluminum Company of America)

stronger at the bend line when the bend takes place across this grain than when it takes place parallel to it. The mechanic should keep this in mind when working aluminum or aluminum alloy sheet. The grain lines run in the direction in which rolling takes place. Figure 330 shows the first drawing operation performed on an aluminum blank in forming a deep-drawn part.

Forming Magnesium Alloys. Most magnesium parts may be formed in one operation, thus avoiding repeated annealing and redrawing. In forming magnesium at elevated temperatures, the deep-drawn parts do not require additional dies. This constitutes a saving in both time and equipment. When formed at approximately 600° F., no allowance need be made for spring back. If worked at lower temperatures, spring-back allowance must be made. For example, Dowmetal J–1H sheet, when

AIRCRAFT MAINTENANCE AND SERVICE

Fig. 331. Airplane wheel fairings formed of magnesium alloy. (Courtesy Dow Chemical Company)

bent around a radius four times the thickness of the metal at 400° F., must have a spring-back allowance of approximately 3°. Table XV shows the most satisfactory forming temperature ranges for magnesium alloys in both the hard-rolled and annealed condition.

TABLE XV. RADII* FOR 90° BENDS IN MAGNESIUM ALLOY SHEET

DOWMETAL ALLOY	CONDITION	MINIMUM RADII AT 70° F.	TYPICAL RADII AT VARIOUS TEMPERATURES			
			70° F.	300° F.	400° F.	600° F.
FS–1a	Annealed	5T**	4½T	—	1.2T	—
FS–1h	Hard rolled	10T	8T	4½–6T	—	—
J–1a	Annealed	9T	7T	—	4–6T	2–3T
J–1h	Hard rolled	17T	13½T	—	4½–6½T	—
Ma	Annealed	7T	5½T	—	3–4T	1–2T
Mh	Hard rolled	12T	9T	—	6–7T	—

* Radii formed in specimens 6 in. wide in square rubber retaining dies.
** T = thickness of sheet.

Magnesium alloys can be hand formed in the same manner as the aluminum alloys. The material must be heated and should be hammered with a leather mallet. In hand forming, it is suggested that metal form blocks and backing blocks be used. The material to be formed is clamped between the blocks, and both the blocks and the metal heated to the desired temperature. The blocks will maintain the

FORMING SHEET METAL

heat of the part being worked for a considerable period of time. A leather mallet is recommended for handworking magnesium although, where necessary, a softwood mallet may be used.

Forming Corrosion-Resistant Alloys. Alloys of steel may be successfully riveted. Small rivets having a diameter of $3/16$ in. or less may be driven cold. Due to the tendency to harden when worked cold, it is recommended that the rivets be set either by a squeeze riveter or by as few blows as possible. A single blow is more desirable than a number of light blows. Larger-sized rivets should be driven while hot. The temperatures at which the rivets are heated should be accurately controlled. The proper temperature is approximately 2000° F. The rivet should be set before the temperature drops below 1600° F. The tem-

Fig. 332. Final forming operation on the stainless steel fitting. (Courtesy The Glenn L. Martin Company)

peratures at which rivets of the stainless-steel alloys are driven vary with the different alloys and should be checked carefully with the manufacturer's specifications. Heating for too long a time tends to cause scale to form on the material. Rivets of the chrome nickel alloys should not be heated for more than approximately 10 min., and the straight chromium alloys should not be heated more than approximately 25 min. Rivet holes may be either drilled or punched. It is necessary to ream

AIRCRAFT MAINTENANCE AND SERVICE

punched holes. Drilled holes are preferred as they cause less strain to the material.

When excessive strains are not to be encountered, stainless-steel alloys may be soldered or brazed. These alloys lend themselves particularly well to spot welding, and this method of fastening is used wher-

Fig. 333. A slab of stainless steel about to enter the rolls. (Courtesy O.W.I. photograph by Al Palmer from a negative now in the Library of Congress)

ever possible. In making repairs to stainless-steel structures, it is not usually possible to spot-weld. It is allowable to use rivets, bolts, or screws. When stainless-steel sheet is not available, repairs may be made by using chromium molybdenum sheet having equal strength.

Inconel, another of the corrosion-resistant alloys, composed of nickel, chromium, and iron, is used in the aircraft industry for exhaust collector rings, manifolds, and other parts subjected to excessive heat. This material is worked in much the same manner as the stainless steels. It should be worked in the annealed condition. Inconel cannot be hardened by heat treatment, but may be readily hardened by cold-working. To anneal Inconel, it should be heated to approximately 1800° F. for about 15 min. The material is then quenched in cold water. When the part has been subjected to excessive cold-working, it should be annealed by the above method to relieve internal stresses. This material does not strain-harden as rapidly as stainless steel. Inconel bends more

Fig. 334. Stainless steel sheet formed by rolling. (O.W.I. photograph by Al Palmer from a negative now in the Library of Congress)

Fig. 335. Finished stainless steel sheet. (O.W.I. photograph by Al Palmer from a negative now in the Library of Congress)

readily than the stainless steels and should bend through an angle of 180° with or against the grain without cracking. The radius of the bending angle should be equal to one half the thickness of the material. Inconel may be welded by the oxy-acetylene, electric arc, spot, or seam process. Parts are usually joined by welding rather than by riveting.

The working characteristics of Monel metal are similar to those of Inconel.

Copper is ductile when in the annealed condition and is easily worked. It becomes hardened through cold-working and must be re-annealed when worked excessively. Copper may be annealed by heating to a red heat, which is approximately 1000° F., and quenching in cold water. Copper parts in direct contact with aluminum alloys should be cadmium plated.

Sheet brass in the annealed condition may be formed readily. When brass parts come in contact with aluminum alloys, they should be cadmium plated. Copper and brass sheets may be joined by soft solder, hard solder, or brazing.

All corrosion-resistant alloys should be worked and treated in accordance with the manufacturer's specifications.

XXII ASSEMBLY

Assembly consists of fastening the various parts and sub-assemblies into place. Several different kinds of fasteners are used. Rivets are one of the most important of the fasteners used in assembly. Riveting is always an assembly process. Some parts may be fastened by means of bolts, screws, self-tapping screws, machine screws, or studs. Many different kind of nuts are used, such as wing nuts, check nuts, self-locking nuts, nut plates, and elastic stop nuts. When bolts are used, they usually have to be safetied by means of cotter pins or safety wire and may be fitted with washers to keep the nut from injuring the structural material. Cowlings are usually fastened into place by special cowling fasteners so that the cowling may be removed or opened for service and inspection. Inspection plates may be fastened with screws which screw into nut plates. Temporary fasteners are often used to hold the parts in place during assembly. These fasteners may consist of clamps, spring clamps, parts of a jig, or Cleco clamps. Many small sub-assemblies are fastened by spot welding, as shown in Figures 336 and 337.

In addition to the various hand tools, assembly requires a number of special jigs. Jigs must be constructed with extreme accuracy. The

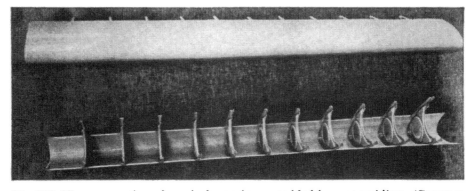

Fig. 336. The nose section of an airplane wing assembled by spot welding. (Courtesy Aluminum Company of America)

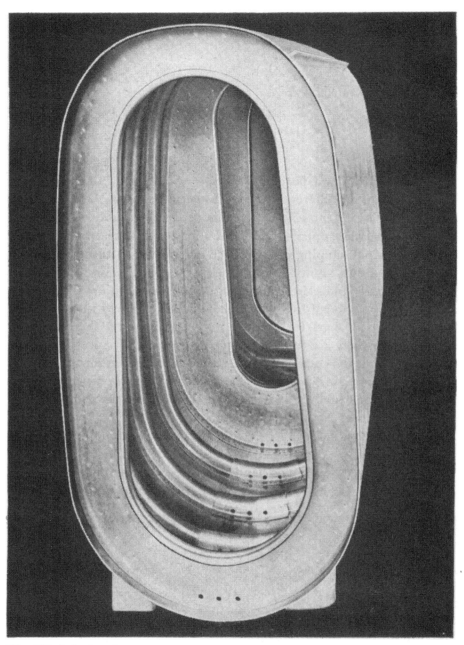

Fig. 337. A fuel tank constructed by spot welding. (Courtesy Aluminum Company of America)

ASSEMBLY

tolerance of most jig structures is not more than $1/100$ in. Jigs vary in size from those used to assemble an aileron, rudder, or fuel tank to those which determine the shape of the entire airplane. These jigs are named after the part which they are designed to support, such as wing

Fig. 338. A wing mounted on a jig for assembly. Note the diagonal rib construction. (Courtesy Engineering Research Corporation)

jigs and fuselage jigs. Some jigs resemble check blocks, over which various parts are assembled.

Jigs should be so formed that the parts fit readily into place and are easily removed when assembly is complete. Many jigs are so formed that the assembly may be rotated to allow easy access to all parts. Figure 338 shows an airplane wing mounted on such a jig. This wing shows diagonal rib construction.

All jigs must be designed for the particular part being assembled. Fuselage jigs are usually heavy, rigid structures designed to hold bulkheads, stringers, cross members, and other structural parts in place while being assembled. Many fuselage jigs are so designed that the skin is fastened to the structural framework while still in the jig.

Light airplanes usually have the wing assembled in one panel. The jig is constructed of two uprights connected with two cross members. The cross members locate the front and rear spars. The fuselage fittings in the root of the wing are usually used for bolting points. The ribs slide over the spars and are fastened into place. Wings are usually as-

Fig. 339. Mass production parts ready for subassembly on an automatic riveting machine. (Courtesy Douglas Aircraft Company)

Fig. 340. Riveting a subassembly. (Courtesy Douglas Aircraft Company)

ASSEMBLY

sembled with the spars horizontal and the wing surfaces vertical. This allows work on both the top and bottom surfaces of the wing at one time. Large wings are built in sections, one of which may consist of a center section. To the center section may be fastened the engine nacelles, engine mounts, and landing-gear parts. Center sections often

Fig. 341. Riveting a wing covering in place. Note the jig and the large number of Cleco clamps holding the skin in place. (Courtesy Cleveland Pneumatic Tool Company)

contain the fuel tanks. The outer portion of the wing may be assembled in two parts consisting of the main panel and the tip. Some large airplanes have a root panel which is fastened directly to the fuselage. Figure 342 shows a fuselage being lowered onto the center section during the assembly of a large, four-motored airplane.

The metal skin of an airplane is usually of thin material which must be carefully handled to prevent damage. The covering should be applied with the grain of the metal extending in the direction of greatest stress. If the grain is difficult to see, it may be brought into view by wiping the sheet with nitrate dope thinner. On wings and control surfaces, the grain should run parallel to the spars. The grain, when pos-

Fig. 342. Wing assembly. These jigs allow the wing to rotate. (Courtesy Cleveland Pneumatic Tool Company)

Fig. 343. Assembling a cabin section. (Courtesy Aluminum Company of America)

ASSEMBLY

sible, should be arranged to run lengthwise on the fuselage. It is the responsibility of the skin fitter to check the grain on all parts of the skin which he applies. Many times a part needs additional forming to prevent the skin from wrinkling or buckling. It may be necessary to remove a part which has been fitted into place and send it back to be reformed. However, the skin fitter should be able to make minor ad-

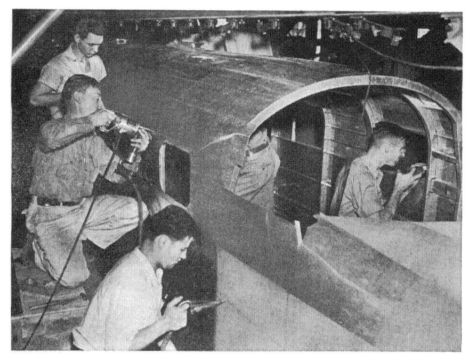

Fig. 344. Fuselage assembly. Note the worker at the right is bucking a rivet; worker at left is drilling rivet holes. (Courtesy Aluminum Company of America)

justments. If the part is undersized, it must be scrapped. When the fit is satisfactory, a helper, working from the inside of the sheet, drills the necessary rivet holes. Pilot holes are usually drilled in the framework before the skin is fitted. These pilot holes in the structural parts to which the skin is to be fastened are usually drilled with a No. 50 drill. No. 30 is a common size drill for skin rivets. Pilot holes may be drilled from the inside of the skin and then reamed from the outside with the proper size of drill before riveting. The portion of skin being drilled is held in place by padded wooden blocks. The first holes are usually drilled near the center of the sheet being applied. As soon as these holes are drilled, skin fasteners, such as Cleco clamps or machine

Fig. 345. Riveting the skin to the fuselage framework. The center section of the stabilizer has been fastened into place. (Courtesy Cleveland Pneumatic Tool Company)

Fig. 346. The assembly line in a large aircraft factory. (Courtesy Douglas Aircraft Company)

ASSEMBLY

screws having fiber washers inserted under the heads to protect the skin, are put in place to secure the sheet. Sheet-metal screws should not be used to hold the sheet in place after the holes have been drilled to full size. The holes are drilled, and the fasteners inserted, working outward from the center of the sheet. Any wrinkles that form must be worked out. Wrinkles and bulges may form oil cans. The skin fitter can test for this condition by pressing on the sheet and determining

Fig. 347. Final assembly. (Courtesy Douglas Aircraft Company)

whether or not it will spring back and forth in the same manner as the bottom of an oil can. Oil cans must always be removed, as vibration may cause failure of the parts.

After the holes are drilled and the skin is properly fitted, all burrs should be removed from the skin and structural members. If the skin is to be flush riveted, any one of a number of different methods may be used. The structural part to which the sheet is riveted may have the holes countersunk or dimpled. Dimples may be formed by a machine or by hand tools. On thin sheet the dimple may be formed by the rivet during the driving operation. The skin is then put in place and tried for proper fit. When the skin fits, the proper number of temporary

Fig. 348. Part of the skin left off of the underside of a wing to allow installation of tubing and acces-

ASSEMBLY

fasteners to hold the sheet firmly in place is inserted through the rivet holes. Riveting is started near the center of the sheet, working outward in the same manner in which the holes were drilled. Any tendency to buckle or deform should be immediately reported to the supervisor. It is often necessary that various parts of the skin be left off temporarily to allow the installation of pipes, conduits, fittings, and accessories.

Fig. 349. Weighing an airplane to determine the center of gravity. (Courtesy Piper Aircraft Corporation)

INDEX

Accessories	75
Acetone	15
Acetylene	15
Aileron	15
Air scoop	15
screw (propeller)	15
Aircraft	7, 15
composite	7
woods	143
Airfoil	15
Airplane	15
Alclad	15
Alignment	15
Alloy	15, 188
Aluminum	15
alloys	15
Angle, attaching	15
of incidence (see Angle of wing setting)	16
of wing setting	16
Annealing	16
Anodize	16
Antidrag wire	16
Assembly	16, 337
Backhand welding	16
Backing strip	16
Bark pockets	163
Base metal (parent metal)	16
Battery	103
Bend allowance	313
chart	323
Bevel	16
closed	16
opened	16
Blowpipe	16
Bonding	16, 55
Bracings	54
Brakes	52, 73
Brashness	16, 166
Brass	336
Brazing	16
Brittleness	16
Broken view	113

Bucking	16
Bulb angle	17
Bulkhead	17
Burring	17, 295, 300
Butt weld	17
Cabin	108
Cables	100
Cadmium	17
Camber	17
Cantilever	17
Carbon tetrachloride	76
Carburizing	17
flame	17
Case hardening	17, 166
Casein glues	118
Center section	17
Chamfer	17
Checks	160
Chord	17
Chromodizing	17
Cleaning	75
Clearance, lip	296
Cloth	17
Cockpit	17
Collapse	164
Components, major	87
Compreg	17
Compression, failure	17
member	17
rib	17
wood	159
Concentric	17
Conductivity	17
Conduit	17
Control, cables	66
pulleys	66
surfaces	11, 17, 64, 103, 107
system	99
Controls	11
Copper	18, 336
Countersink	18
Cowling	11, 18, 52, 74
Cross section	112

349

INDEX

Cutting, blowpipe	18
tip	18
Cylinder (bottle)	18
Datum line	18
Defects, woods	152
Deposited metal	18
Density	18
Die casting	18
Dies	18
Dimpling	18
Distortion	18
Dope	18
Dopes	129
Dowel	18
Drag wire	18
Drill	18
sizes	299
speed	299
twist	294
Drilling	295
Durability	18
Dural	18
Edge distance	18
Elastic limit	19
Elasticity	18
Electrolytic	19
Elongation	19
Empennage	11, 19, 42
Engine	106
controls	104
mount	11, 60
mounts	45
starting	83
Erection procedure	80
Exhaust collector ring	19
Expansion coefficient	19
Extrusion	19
Fabric	19
covered construction	201
Fairing	19, 52
Fairings	74
Fatigue	19
Ferrous	19
Ferrule	19
File	305
Filing	295, 304
Filler or welding rod	19
Fillet weld	19
Filter lens	19
Fin	19
Final assembly	19
Finishes	129
Finishing	101
Firewall	19
Fitting	19
Fixed surfaces	107
Flange	19
Flap	19
Flaps	13
Flat pattern	320
Flight-control mechanism	53
Flux	20
Forging	20
Former or false rib	20
Forming	329
Framework	20
Fuel system	95
Fuselage	9, 20, 34, 60, 90, 107
Gauge	20
Glue, casein	20
Glues	118
synthetic resin	124
Grain	20, 152
Gusset	21
Hardness	21
tests for metal	21
Hardwoods	21
Heart, heartwood	21
Heat treatment	21
of aluminum alloys	21
Honeycomb	21
Honeycombing	164
Horn	21
Hot short	21
Icebox rivet	307
Inconel	21, 334
Inspection	28, 30
after-flight	106
daily	106
hole	21
preflight	102
service	102
Instruments	99, 104
Jig	21
Joint	21
splice	21
Kerf	21
Kiln	21
Knot	21
Knots	156

INDEX

Laminated wood	21
Laminations	127
Landing gear	11, 46, 66, 92, 103, 107
Layout	313
Leading edge	22
Lightening hole	22
Lip clearance	296
Longeron	22
Loom	22
Lucite	22
Lumber	22
Magnesium alloys	331
Magnetic inspection	22
Main beam	22
Maintenance	30
general	83
Manifold	22
Masonite	22
Melting point	22
Member	22
Metal	22
Metals	170
aircraft	170
Moisture content of wood	22
Mold	22
line	22
Monel	336
metal	23
Monocoque fuselage	23
Motor mount	23
Nacelle	23
Nacelles	45
Neutral flame	23
Nose	23
Oil system	103
Oilcan (colloquial)	23
Overhaul	30
Oxide	23
Oxidizing flame	23
Oxy-acetylene welding	23
Oxygen	23
Peening	23
Penetration	23
Pickling	23
Pictorial drawing	110
Pitch, pocket	23
pockets	160
streaks	163
Plastic	23
Plexiglas	24

Ply	24
Plywood	24, 118, 127
Porosity	24
Power, brake	24
plant	13
Preheating	24
Primer	24
Propeller	103, 106
Ream	24
Repair	30
Repairs, approved	135
emergency	136
sheet metal	310
welded	137, 265
Rib	24
box	24
form	24
ordinary	24
Rig (airplane)	24
Rigging	80
Ring, annual-growth	24
cowling	24
Ripple	24
Rivet, draw	24
gauge	24
heads	307
set	24
Rivets	308
Riveting	24, 295, 306
Riveting, blind	24
gun	24
hand	24
Safety practices	114
Salt water	76
Sap	25
Sapwood	25
Scarf joints	123
Seasoning	25
Set back chart	321
Shake	25
Shakes	160
Sheet metal forming	326
Shock-cord	93
Shrinking	326
Skip welding	25
Slag	25
Softwoods	25
Solder	25
Span	25
Spar, main	25
rear	25
Spark test	171

351

INDEX

Specifications	113
Specific gravity	25
Spinner	25
Splice	25
Split	25
Splits	160
Spoilers	13
Spot welding	25
Spring wood	25
Stainless steel	25, 334
Strain	26
Stress	26
Stretching	326
Strut	26
fuselage	26
Subassembly	26
Summer wood	26
Surfaces, flight-control	42
Synthetic resin glues	124
Tack weld	26
Tail, group	11, 42, 90
wheel	103
Tailboom	26
Tap	26
Technical drawings	109
Temperature data	173
Template	26
Tensile strength	26
Three-view drawing	110
Tires	71
Tip (blowpipe)	26
Trailing edge	26
Transparencies	75
Undercut	26
Upkeep	30
Upsetting	26
Veneer	26
Wane	26
Warm-up, engine	106
Warp	26
Web	26
Welded repairs	265
Welding hose	26
Wheels	50, 72
Wing	26
panel	87
rib	27
struts	54
tip	27
Wings	9, 39, 60
Wood, compression	159
defects	168
Woods, aircraft	143
Workability	27
Yield strength	27

The Aviation Collection by Sportsman's Vintage Press

www.SportsmansVintagePress.com

Aircraft Construction Handbook	by Thomas A. Dickinson
Aircraft Sheet Metal Work	by C. A. LeMaster
The Aircraft Apprentice	by Leslie MacGregor
Aircraft Woodwork	by Col. R. H. Drake
Aircraft Welding	by Col. R. H. Drake
Aircraft Sheet Metal	by Col. R. H. Drake
Aircraft Engines	by Col. R. H. Drake
Aircraft Electrical and Hydraulic Systems, and Aircraft Instruments	by Col. R. H. Drake
Aircraft Engine Maintenance and Service	by Col. R. H. Drake
Aircraft Maintenance and Service	by Col. R. H. Drake

Printed in Poland
by Amazon Fulfillment
Poland Sp. z o.o., Wrocław